T0274917

NEW LAWS OF ROBOTICS

NEW LAWS OF ROBOTICS

DEFENDING HUMAN EXPERTISE IN THE AGE OF AI

FRANK PASQUALE

THE BELKNAP PRESS *of* HARVARD UNIVERSITY PRESS

Cambridge, Massachusetts & London, England   2020

First printing

Library of Congress Cataloging-in-Publication Data

Names: Pasquale, Frank, author.
Title: New laws of robotics : defending human expertise in the age of AI /
    Frank Pasquale.
Description: Cambridge, Massachusetts : The Belknap Press of Harvard
    University Press, 2020. | Includes bibliographical references and index.
Identifiers: LCCN 2020012258 | ISBN 9780674975224 (cloth)
Subjects: LCSH: Robotics—Law and legislation. | Artificial intelligence—
    Law and legislation. | Virtual humans (Artificial intelligence)
Classification: LCC K564.C6 P375 2020 | DDC 343.09 / 99—dc23
LC record available at https://lccn.loc.gov/2020012258

To my friends and colleagues at the Association for the Promotion of Political Economy and the Law—a true intellectual community.

And, of course, to Ray.

CONTENTS

Education is the point at which we decide whether we love the world enough to assume responsibility for it, and by the same token save it from that ruin which, except for renewal, except for the coming of the new and young, would be inevitable. And education, too, is where we decide whether we love our children enough not to expel them from our world and leave them to their own devices, nor to strike from their hands their chance of undertaking something new, something unforeseen by us, but to prepare them in advance for the task of renewing a common world.

—*Hannah Arendt,* Between Past and Future

I am speaking of a law, now, understand,
that point at which bodies locked in cages
become ontology, the point at which
structures of cruelty, force, war,
become ontology. The analog
is what I believe in, the reconstruction
of the phenomenology of perception
not according to a machine,
more, now, for the imagination to affix to
than ever before.

—*Lawrence Joseph, "In Parentheses"*

NEW LAWS OF ROBOTICS

1

▪ Introduction

The stakes of technological advance rise daily. Combine facial recognition databases with ever-cheapening micro-drones, and you have an anonymous global assassination force of unprecedented precision and lethality. What can kill can also cure; robots could vastly expand access to medicine if we invested more in researching and developing them. Businesses are taking thousands of small steps toward automating hiring, customer service, and even management. All these developments change the balance between machines and humans in the ordering of our daily lives.

Avoiding the worst outcomes in the artificial intelligence (AI) revolution while capitalizing on its potential will depend on our ability to cultivate wisdom about this balance. To that end, this book advances three arguments that stand to improve all our lives. The first is empirical: right now, AI and robotics most often complement, rather than replace, human labor. The second proposes a value: in many areas, we should maintain this status quo. And the final point is a political judgment: our institutions of governance are actually capable of achieving exactly that outcome. Here is this book's most basic premise: we now have the means to channel technologies of automation, rather than being captured or transformed by them.

These ideas will strike many as commonsensical. Why write an entire book to defend them? Because they have some surprising implications, which should change how we organize social cooperation and deal with conflict. For example, at present, too many economies favor capital over labor and consumers over producers. If we want a just and sustainable society, we must correct these biases.

That correction will not be easy. Ubiquitous management consultants tell a simple story about the future of work: if a machine can record and imitate what you do, you will be replaced by it.[1] A narrative of mass unemployment now grips policymakers. It envisions human workers rendered superfluous by ever-more-powerful software, robots, and predictive analytics. With enough cameras and sensors, this story goes, managers can simulate your "data double"—a hologram or robot that performs your job just as well, at a fraction of your wages. This vision offers stark alternatives: make robots, or get replaced by them.[2]

Another story is possible and, indeed, more plausible. In virtually every walk of life, robotic systems can make labor more valuable, not less. This book tells the story of doctors, nurses, teachers, home health aides, journalists, and others who work *with* roboticists and computer scientists, rather than meekly serving as data sources for their future replacements. Their cooperative relationships prefigure the kind of technological advance that could bring better health care, education, and more to all of us, while maintaining meaningful work. They also show how law and public policy can help us achieve peace and inclusive prosperity rather than a "race against the machines."[3] But we can only do so if we update the laws of robotics that guide our vision of technological progress.

ASIMOV'S LAWS OF ROBOTICS

In the 1942 short story "Runaround," the science fiction writer Isaac Asimov delineated three laws for machines that could sense their environment, process information, and then act.[4] The story introduces a "Handbook of Robotics, 56th Edition," from 2058, that commands:

1. A robot may not injure a human being or, through inaction, allow a human being to come to harm.
2. A robot must obey the orders given it by human beings except where such orders would conflict with the First Law.
3. A robot must protect its own existence as long as such protection does not conflict with the First or Second Laws.

Asimov's laws of robotics have been enormously influential. They seem clear-cut, but they are not easy to apply. Can an autonomous drone blast a terrorist cell? The first half of the first law ("A robot may not injure a human being") seems to forbid such an action. But a soldier might quickly invoke the second half of the first law (forbidding "inaction" that would "allow a human being to come to harm"). To determine which half of the law should apply, we must look to other values.[5]

The ambiguities do not stop at the battlefield. Consider, for instance, whether Asimov's laws allow robotic cars. Self-driving vehicles promise to eliminate many thousands of traffic fatalities each year. So the problem may seem easy at first glance. On the other hand, they would also put hundreds of thousands of paid drivers out of work. Does that harm entitle governments to ban or slow down the adoption of self-driving cars? Asimov's three laws are not clear on such matters. Nor do they have much to say about a recent demand of self-driving car evangelists: that pedestrians be trained to act in ways that make it easier for self-driving vehicles to operate, and penalized if they fail to do so.

These ambiguities and many more are why the statutes, regulations, and court cases affecting robotics and AI in our world are finer grained than Asimov's laws. We will explore much of that legal landscape in the course of the book. But before doing so, I want to introduce four new laws of robotics to animate our explorations ahead.[6] They are directed to the people building robots, not to robots themselves.[7] And although they are more ambiguous than Asimov's, they also better reflect how actual law-making is accomplished. Because legislators cannot possibly anticipate every situation that authorities may need to address, they often empower agencies with broadly worded statutes. New laws of robotics should be similar, articulating broad principles while delegating specific authority to dedicated regulators with long experience in technical fields.[8]

NEW LAWS OF ROBOTICS

With these goals in mind, four new laws of robotics will be explored and advanced in this book:

1. **Robotic systems and AI should complement professionals, not replace them.**[9]

Clashing projections of technological unemployment drive popular discussions of the future of work. Some experts predict that almost every job is destined to be whittled away by technological advance. Others point out roadblocks on the path to automation. The question for policymakers is, Which of these barriers to robotization make sense, and which deserve scrutiny and removal? Robotic meat-cutters make sense; robotic day care gives us pause. Is the latter caution mere Luddite reaction, or is it reflective of a deeper wisdom about the nature of childhood? Licensure rules now keep symptom-analyzing apps from being marketed as physicians practicing medicine. Is this good policy?

This book analyzes such examples and marshals both empirical and normative arguments for slowing down or speeding up the adoption of AI in varied fields. Numerous factors matter, specific to jobs and jurisdictions. But one unitary, organizing principle is the importance of meaningful work to the self-worth of people and the governance of communities. A humane agenda for automation would prioritize innovations that complement workers in jobs that are, or ought to be, fulfilling vocations. It would substitute machines to do dangerous or degrading work, while ensuring those presently doing that work are fairly compensated for their labor and offered a transition to other social roles.

This balanced stance will disappoint both technophiles and technophobes. So, too, will the emphasis on governance alienate both those who object to "interference" in labor markets, and others who loathe a "professional-managerial class." To the extent professions amount to an economic caste system, unfairly privileging some workers over others, their suspicions are justified. However, it is possible to soften stratification while promoting professions' higher aims.

The bargain at the core of professionalism is to empower workers to have some say in the organization of production, while imposing duties upon them to advance the common good.[10] By advancing research, whether in divisions of universities or in their own offices, professionals cultivate *distributed expertise*, alleviating classic tensions between technocracy and popular rule. We should not be dismantling or disabling professions, as all too many advocates of disruptive innovation aspire to do. Rather, humane automation will require the strengthening of existing communities of expertise and the creation of new ones.

A good definition of professions is capacious, and it should include many unionized workers, particularly when they protect those they serve from unwise or dangerous technologies. For instance, teachers' unions have protested excessive "drilling and testing" via automated systems and have promoted their students' interests in many other contexts. Unions that tend toward a path of professionalization—empowering their members to protect those they serve—should have an important role in shaping the AI revolution.

Sometimes it will be difficult to demonstrate that a human-centered process is better than an automated one. Crude monetary metrics crowd out complex critical standards. For example, machine learning programs may soon predict, based on brute-force natural language processing, whether one book proposal is more likely than another to be a best seller. From a purely economic perspective, such programs may be better than editors or directors at picking manuscripts or film scripts. Nevertheless, those in creative industries should stand up for their connoisseurship. Editors have an important role in publishing, exercising judgment, and finding and promoting work that the public may not (now) want, but needs. The same could be said of journalists; even if automated text generation could generate ad-maximizing copy, that hollow triumph should never replace genuine reporting from an authentic, hard-won, human point of view. Professional schools in universities clarify and reexamine standards in media, law, medicine, and many other fields, preventing them from collapsing into metrics simple enough to be automated.

Even in fields that seem most subject to the automation imperative, in areas like logistics, cleaning, agriculture, and mining, workers will play a critical role in a long transition to AI and robotics. Gathering or creating the data necessary for AI will be a demanding task for many. Regulations can make their jobs more rewarding and self-directed. For example, European privacy law empowers drivers to resist the kind of 360-degree surveillance and control that oppresses truckers in the United States.[11] That is not to say that such a dangerous occupation should go unmonitored. Sensors may indicate problems with a driver's reflexes. But there is a world of difference between sensors specifically aimed at safety lapses, and a constant video and audio recording of all actions. Getting the balance right between unnerving and demeaning surveillance and sensible, focused monitoring will be crucial in a wide range of fields.

We can also design technological transitions that keep human beings in the picture, or at least give them that choice. For example, Toyota has promoted cars with a spectrum of machine involvement, from chauffeur mode (which requires minimal monitoring by a driver) to guardian mode (which focuses the car's computing systems on accident avoidance, while a person helms the vehicle).[12] Planes have had autopilot capacities for decades, but commercial carriers still tend to have at least two people in the cockpit. Even the occasional airline passenger can be grateful that the evangelists of substitutive automation are in no hurry to jettison pilots.[13]

Note, too, that transportation is one of the easier cases for AI. Once a destination is set, there is no argument over the point of a trip. In other service fields, the opposite is true: the customer's or client's mind may change. A class may be too antsy on a beautiful spring day to practice multiplication tables repeatedly. A socialite may call his interior designer, worried that the print they chose for the living room is too brash. A trainer may vacillate, worried that her client is too exhausted to run another minute on a treadmill. In each of these cases, communication is key, as are the human-to-human skills of patience, deliberation, and discernment.[14]

Yes, if thousands of trainers equipped themselves with Google Glass and recorded all their encounters, perhaps some divine database of grimaces and rolled eyes, injuries, and triumphs could dictate the optimal response to a miserable gym goer. But even to begin imagining how to construct such a database—what gets marked as a good or bad outcome, and to what extent—is to understand the critical role people will play in constructing and maintaining a plausible future of AI and robotics. Artificial intelligence will remain artificial because it will always be a product constructed out of human cooperation.[15] Moreover, most recent advances in AI are designed to perform specific tasks rather than to take on entire jobs or social roles.[16]

There are many examples of technologies that make jobs more productive, more rewarding, or both. As the Agency for Digital Italy has observed, "Technology often does not completely replace a professional figure but replaces only some specific activities."[17] Contemporary law students can barely believe that pre-internet lawyers had to comb through dusty tomes to assess the validity of a case; research software makes that process easier and vastly expands the range of resources available for an argument. Far from simplifying matters, it may make them much more complex.[18] Spending less time hunting down books and more time doing the intellectual work of synthesizing cases is a net plus for attorneys. Automation can bring similar efficiencies to myriad other workers, without mass displacement of labor. And this is not merely an observation. This is a proper goal of policy.[19]

2. Robotic systems and AI should not counterfeit humanity.

From Asimov's time to the vertiginous mimicry of *Westworld,* the prospect of humanoid robots has been fascinating, frightening, and titillating. Some roboticists aspire to find the right mix of metal bones and plastic skin that can break out of the "uncanny valley"—the queasiness that a humanoid robot evokes when it comes very close to, but does not quite recreate, human features, gestures, and ways of being. Machine-learning programs have already mastered the art of creating pictures of "fake people," and convincing synthetic voices may soon become common.[20] As engineers scramble to fine-tune these algorithms, a larger question goes unasked: Do we want to live in a world where human beings do not know whether they are dealing with a fellow human or a machine?

There is a critical difference between humanizing technology and the counterfeiting of distinctively human characteristics. Leading European ethicists have argued that "there have to be (legal) limits to the ways in which people can be led to believe that they are dealing with human beings while in fact they are dealing with algorithms and smart machines."[21] Lawmakers have already passed "bot disclosure" laws in online contexts.

Despite this growing ethical consensus, there are subfields of AI—such as affective computing, which analyzes and simulates human emotion—devoted to making it more and more difficult for us to distinguish between humans and machines. These research projects might culminate in a creation like the advanced androids in the Steven Spielberg film *A.I.*, indistinguishable from a human being. Ethicists debate how such humanoid robots should be designed. But what if they should not be made at all?

In hospitals, schools, police stations, and even manufacturing facilities, there is little to gain by embodying software in humanoid bodies, and plenty to lose. The race to mimic humanity might too easily become a prelude to replacing it. Some people might prefer such replacement in private life, and law should respect such autonomy in intimate realms. But the idea of a society dedicated to advancing it in workplaces, the public sphere, and beyond is madness. It confuses the abolition of humanity with its advance.

This argument may jar or confound technophiles—to reject not merely the substance, but also the premise, not only of Asimov's laws, but also of a vast literature on the future of technology. I hope to justify this conservatism by thinking through, chapter by chapter, the concrete steps we would need to take to get to a science-fictional world of robots indistinguishable from humans. That transition entails massive surveillance of humans, all to create robots designed to fool or allure human beings into treating machines as their equals. Neither prospect is appealing.

The voice or face of another human being demands respect and concern; machines have no such claim on our conscience. When chatbots fool the unwary into thinking that they are interacting with humans, their programmers act as counterfeiters, falsifying features of actual human existence to increase the status of their machines. When the counterfeiting of money reaches a critical mass, genuine currency loses value. Much the same fate lies in store for human relationships in societies that allow machines to freely mimic the emotions, speech, and appearance of humans.

The counterfeiting of humanity is a particular danger as corporations and governments seek to put a friendly face on their services and demands. Google Assistants have wowed the business press by mimicking secretaries making appointments, eerily replicating even the "ums" and "ahs" that punctuate typical phone conversations. These conversational fillers disguise the power of a firm like Google with the hesitation or deference typically expressed by a human's unpolished speech. They cloak a robocall as a human inquiry. For those on the receiving end of the calls, it is all too easy to imagine abuse: a deluge of calls from robotized call centers.

Counterfeiting humanity is not merely deceptive, it is also unfair, giving the counterfeiter the benefit of the appearance of personal support and interest without its reality. As we will see in case after case— of robot teachers, soldiers, customer-service representatives, and more—dissatisfaction and distress at failed imitations of humanity are not merely the result of imperfect technology. Rather, they reflect wise caution about the direction of technology itself.

3. Robotic systems and AI should not intensify zero-sum arms races.

Debates over "killer robots" are a central theater for ethics in international law. A global coalition of civil society organizations is pushing nations to pledge not to develop lethal autonomous weapons systems (LAWS). Several factors now stymie this commendable proposal for technological restraint. Military leaders distrust their counterparts in rival countries. They may hide militarized AI research, advancing in power even as they publicly disclaim any such intent. Rising powers may assert themselves, investing in force projection to match their new economic status, while now-dominant militaries press for more resources to maintain their relative advantage. This is but one of many ways an arms race begins. As AI and robotics enter the picture, the stakes of falling behind one's rivals rise, since emerging technologies promise to be so much more targeted, ubiquitous, and rapidly deployed.

Dovish politicians may commit themselves to a purely defensive posture (reflected in the United States' shift from a Department of War to a Department of Defense in 1949). But defenses can often be repurposed as offensive weapons; think, for instance, of autonomous drones designed to destroy missiles but reprogrammed to assassinate generals. Thus, even protective plans can seem aggressive, as in the case of Ronald Reagan's Strategic Defense Initiative (SDI). Popularly known as Star Wars, SDI would have relied on lasers in space to shoot down Soviet missiles. Had it worked, it would have upset a fragile balance of deterrence (mutually assured destruction via nuclear annihilation). Now, LAWS, automated cyberattacks, and disinformation campaigns threaten to disrupt long-settled expectations about the purpose and limits of international conflict. We must find new ways of limiting their development and impact.

War may at first appear as a state of exception, where ordinary ethical reasoning is suspended (or at least radically circumscribed). But the third new law of robotics has applications well beyond the battlefield. Technologies pioneered by armies also tempt police departments, and more law enforcers aim to use facial recognition to scan crowds for criminals. Via machine learning, tax authorities may ferret out unreported income by parsing the email and bank accounts of all citizens. The mere possibility of such perfected surveillance will provoke the more security conscious to invest in encryption, which in turn will prod authorities to throw more resources into decryption.

We need to find ways of limiting these dynamics—and not merely in military and policing scenarios. Investment in AI and robotics is often part of a contest for a fixed set of resources—for example, in litigation, finance, and other sectors where people compete for positional advantage. To allocate such resources, government and corporations set citizens and consumers in reputational competitions, such as credit scoring, whose ratings are meaningful to the extent that they stigmatize some people (with low scores) and elevate others (with high ones). At its beginning, credit scoring was confined to one area of life (determining eligibility for loans) and based on a limited set of data (repayment history). Over decades, credit scores and similar measures have come to inform other determinations, including insurance rates and employment opportunities. More recently, data scientists have proposed more data sources for credit scoring, ranging from the way people type, to their political affiliation, to the types of websites they visit online. The Chinese government has also expanded the stakes of surveillance, proposing that "social credit scores" play a role in determining what trains or planes a citizen can board, what hotels a person can stay in, and what schools a family's children can attend. And many of its jurisdictions have expanded the data potentially included—everything from how a person crosses the street, to how they treat their parents, to their degree of patriotism and loyalty to the Communist Party and its teachings.

Social credit scoring and its many Western analogues have been enormously controversial, and it is unclear how far they will develop. To be sure, some applications of these systems may be quite valuable. It is hard to complain about public health surveillance that accelerates contact tracing to stop the spread of infectious disease before a pandemic starts. But when the same powerful capacities are ranking and rating everyone all the time, they become oppressive.

The chief danger of social control via AI is a world of regimentation. Conflict and competition are part of life, and we should expect technological advance to inform them. But AI and robotics threaten to make social control *too* perfect, and competition (to be the entity capable of imposing or evading that control) too fierce. Human security and creativity thrive in environments that balance predictability and openness, order and flux. If we fail to limit robotics embedded in systems of social control, that balance will be upended.

4. Robotic systems and AI must always indicate the identity of their creator(s), controller(s), and owner(s).

People are responsible for robotic or algorithmic systems. True, some programs can now generate new programs, which in turn may spawn others. But we can still trace these "mind children" and grandchildren back to their source.[22] We should keep this status quo for the foreseeable future, despite predictable resistance from some advocates of fully autonomous AI.

The cutting edge of the AI, machine-learning, and robotics fields emphasizes autonomy, whether of smart contracts, high-frequency trading algorithms (at least in time spans undetectable by humans), or future robots. There is a nebulous notion of "out of control" robots that escape their creator. Perhaps such accidents are unavoidable. Nevertheless, some person or entity should be responsible for them. A requirement that any AI or robotics system has some designated party responsible for its action would help squelch such projects, which could be just as dangerous as unregulated bioengineering of viruses.

Of course, some robots and algorithms will evolve away from the ideals programmed into them by their owners as a result of interactions with other persons and machines (think, for instance, of advanced self-driving cars that evolve as a result of multiple influences).[23] In such cases, there may be multiple potentially responsible parties for any given machine's development and eventual actions.[24] Whatever affects the evolution of such machines, the original creator should be obliged to build in certain constraints on the code's evolution to both record influences and prevent bad outcomes. Once another person or entity hacks or disables those constraints, the hacker is responsible for the robot's wrongdoing.

For a concrete application of this principle, consider a chatbot that gradually learns certain patterns of dialogue from interactions on Twitter. According to some news accounts, Microsoft's AI chatbot, Tay, quickly adopted the speech patterns of an unhinged Nazi sympathizer after only a few hours on Twitter.[25] Microsoft did not program that outcome, but it should have known that it was a danger of exposing a bot to a platform notorious for its poor moderation of harassment and hate speech. Moreover, to the extent that the chatbot did log where the malign influences came from, it could have reported them to Twitter, which, in some better version of itself, could have taken some action to suspend or slow the flood of abuse coming from troll accounts and worse.

Regulators will need to require responsibility-by-design (to complement extant models of security-by-design and privacy-by-design). That may involve requiring certain hard-coded audit logs, or licensing practices that explicitly contemplate problematic outcomes.[26] Such initiatives will not simply regulate robotics and AI *post hoc*, but will influence systems development by foreclosing some design options and encouraging others.[27]

. . .

Each of these new laws of robotics, promoting complementarity, authenticity, cooperation, and attribution, rests on a theme that will animate the rest of our exploration: the critical distinction between technology that replaces people and technology that helps them do their jobs better. The point of the new laws is to develop policies that capitalize on human strengths in fields such as health and education, and to take advantage of human limits to bound the scope and intensity of conflict and regimentation in social life.

AI researchers have long aimed to create computers that can sense, think, and act like humans. As far back as the 1960s, roboticists at MIT were developing robot sentries to relieve soldiers of the boring and dangerous duty of standing guard at vulnerable sites.[28] But there is another way to think about the sentry robot—not as AI replacing troops, but as one more tool to increase soldiers' effectiveness as defenders. An army does not necessarily need to requisition more soldiers to monitor emerging threats. It can instead develop sensors and computers designed to act as a second set of eyes and ears, rapidly processing threat levels and other data to better inform soldiers' actions. This goal, deemed "intelligence augmentation" (IA), has informed the projects of many internet pioneers.[29] It is also a mainstay of modern warfare, as drone pilots handle a rich array of sensor data to make life-and-death decisions about aerial bombings.

Though sometimes blurry, the distinction between AI and IA is a critical one for innovation policy. Most parents are not ready to send their children off to robot teachers. Nor should their children be taught that teachers will eventually be replaced by machines perfectly personalized to their learning styles. There are many visions of robotics in education that are far more humane. For example, schools have already experimented successfully with "companion robots" that help young students by drilling them on vocabulary words and asking them questions about what they just learned. Looking like animals or fanciful creatures rather than people, these robots do not challenge the distinctiveness of humanity.

Researchers are finding that in many contexts, IA results in better service and outcomes than either artificial or human intelligence working alone. Assistive AI and robotics could be a godsend for workers, freeing more hours for rest or leisure. But in any modern market economy, there are economic laws that tilt the scale toward AI and against IA. A robot does not demand time off, a fair wage, or health insurance. When labor is viewed primarily as a cost, its fair pay is a problem, which machines are supposed to solve. Robotics revolutionized manufacturing by replacing assembly-line workers. Now, many business experts want similar technological advances to take over more complex work, from medicine to the military.

Caught up in these managerialist enthusiasms, too many journalists have discussed "robot lawyers" and "robot doctors" as if they are already here. This book will show that such descriptions are overblown. To the extent that technology transforms professions, it has tended to do so via IA, not AI. Submerged beneath breathless headlines about "software eating the world," there are dozens of less spectacular instances of computation helping attorneys or doctors or educators to work faster or better.[30] The question now for innovation policy is where to sustain this predominance of IA, and where to promote AI. This is a problem we must confront sector by sector, rather than hoping to impose a one-size-fits-all model of technological advance.

Conversations about robots usually tend toward the utopian ("machines will do all the dirty, dangerous, or difficult work") or the dystopian ("and all the rest besides, creating mass unemployment"). But the future of automation in the workplace—and well beyond—will hinge on millions of small decisions about how to develop AI. How far should machines be entrusted to take over tasks once performed by humans? What is gained and lost when they do so? What is the optimal mix of robotic and human interaction? And how do various rules—from codes of professional ethics to insurance policies to statutes—influence the scope and pace of robotization in our daily life? Answers to these questions can substantially determine whether automation promises a robot revolution or a slow, careful improvement in how work is done.

. . .

Why should we be especially concerned with robots and AI, as opposed to the ubiquitous screens and software that have already colonized so much of our time? There are two practical reasons. First, the physical presence of a robot can be far more intrusive than any tablet, smartphone, or sensor; indeed, those technologies can simply be embedded into robots.[31] No flat-screen could reach out and restrain a misbehaving child or a defiant prisoner, modifying and repurposing present technology of crowd control for new forms of discipline. But a robot could.

Even if uptake of robots is slow or limited, AI threatens to super-charge the techniques of fascination and persuasion embedded into tech ranging from mobile apps to video poker.[32] As human-computer interaction researcher Julie Carpenter observes, "Even if you know a robot has very little autonomy, when something moves in your space and it seems to have a sense of purpose, we associate that with some-thing having an inner awareness or goals."[33] Even something with as little animation as a robot vacuum cleaner can provoke an emotional response. The more sensors record our reactions, the richer the veins of emotional data for more sophisticated computers to mine.[34] Every "like" is a clue to what engages us; every lingering moment on a screen is positive reinforcement for some database of manipulation. Minia-turized sensors make surveillance mobile, unraveling efforts to hide. Indeed, the decision to shield oneself from sensors may be one of the most revealing activities one can engage in. Moreover, processing ca-pacities and data storage could put us on a path to a dystopia where everything counts, and anything a student does may be recorded and backed up to inform future evaluations.[35] By contrast, a student in an ordinary school may encounter different teachers every year, starting off with a relatively clean slate each time.[36]

None of these troubling possibilities is destined to happen, though, which raises a second reason to focus on robotics policy now. As robots enter highly regulated fields, we have a golden opportunity to shape their development with thoughtful legal standards for privacy and consumer protection. We can channel technology through law.[37] Robots need not be designed to record every moment of those whom they accompany or supervise. Indeed, robot supervision itself could seem sufficiently op-pressive that we require some human monitoring of any such system (as a robotic South Korean prison mandated for its mechanical guards). When robots are part of a penal system, a broad, rich debate on prison policy and on the relative merits of retribution and rehabilitation should inform any decision to deploy them. Indeed, one of the main points of the new laws of robotics is to warn policymakers away from framing controversies in AI and robotics as part of a blandly general "technology policy," and toward deep engagement with domain experts charged with protecting important values in well-established fields.

Cynics may scoff that such values are inherently subjective and that they are destined for obsolescence in an ever more technological society. But communities of scholars and consultants focused on science, technology, and human values have shown that anticipatory ethics can inform and influence technology design.[38] Values are designed into technology.[39] Canadian, European, and American regulators have already endorsed privacy-by-design as a basic principle for developers.[40] Rules like that should apply a fortiori to sensor-laden technology, which can move freely to maximize its ability to record images and sound. For example, in the same way that many video cameras have a red light indicating that they are recording, robots should feature an equivalent indicator when they record persons around them. AI-driven data should be subject to strict limits on collection, analysis, and use.[41]

Technologists may counter that it is too early to regulate robotics. Let problems develop and only then move to counter them, say the partisans of laissez-faire. But quietism misses the mark. All too often in high technology fields, industry says it is *never* a good time to regulate. When troubling new business practices emerge, would-be regulators are accused of strangling an "infant industry." Once the practices are widespread, the very fact of their ubiquity is offered as proof that consumers accept them. For every argument offered for legal intervention, there is a pat strategy of deflection based on bromides and platitudes. "Is there really a problem?," "Let's wait and see," and "Consumers want it" are all offered as all-purpose rationales for inaction, played like trump cards in a game of whist.[42]

A wait-and-see attitude ignores the ways in which technology, far from being independent of our values, comes to shape them.[43] Robotic companions for children in online charter schools would not merely reflect (or distort) current values regarding what kinds of socialization are owed to the young. They would also shape those values of those generations, inculcating them with a sense of what types of moments are private and what are fair game for potentially permanent recording. These mores should not simply result from whatever is most profitable for edtech providers. They need democratic governance and input from experts well outside technology fields.[44]

.   .   .

This value-shaping role of technology is also a clear and present danger in warfare, where robotics could fundamentally alter what we take to be the parameters of a fair conflict. For some futurists, the automation of conflict is a foregone conclusion. No military power can afford to fall too far behind rivals that are developing a fearsome fleet of "killer robots."[45] If we are "wired for war," then we are prone to escalate the development of deadly robotic force.[46] On these accounts, human nature dictates a certain course of technological development—and perhaps its own obsolescence—in favor of superhuman robotic systems.[47]

Such realism may be prudent, but it also risks becoming a dangerously self-fulfilling prophecy, accelerating rather than predicting an arms race. The less costly military intervention appears, the more likely politicians and states are to engage in it. Moreover, the more precisely force can be deployed, the more the language of war and law enforcement starts to blur and create ethical gray areas. Consider a near-real possibility: the United States supplements its aerial drone presence in combat zones with ground-crawling robots and smaller, indoor drones. Would the individuals tracked down by such robots be combatants or suspects? Both international and national legal precedents dictate different treatment for each. Such treatment cannot be readily automated—if indeed it can be automated at all. Thus, the law of war (or mere criminal procedure) may prove an insuperable barrier to robot soldiers—or at least their legitimate deployment.[48]

Both academics and government officials have already begun to analyze scenarios of robotized war and law enforcement.[49] Expect a continuing merger of the two fields under the catch-all rubric of "homeland security"—and many more applications of robotics in the name of social order. The promise here is ultimately one of superhuman intelligence in threat detection: AI able to parse millions of data streams in order to quickly detect and defuse future crime.

Before anyone gets too enthusiastic about this technology, however, there are some sobering examples of just how inhuman an artificial

intelligence can be. Researchers have used machine learning to predict criminality from little more than a person's face. Should future
police robots include this facial criminality data as they make decisions about which individuals to follow more closely and which to ignore? Are any data or inferences off limits to machines that may be an
increasing part of law enforcement in the future, such as the predictive
analytics now popular in some police departments? Do we have the
right to inspect and dispute such data? Should such research even be
done at all?[50]

Correlating facial features with criminality may seem like an exotic or
unusual application of AI. But the underlying logic of much advanced
computation remains impenetrable to ordinary forms of explanation.
Some roboticists celebrate inexplicability as an alternative to—or even
advance beyond—human intelligence. "At some point, it's like explaining
Shakespeare to a dog," Columbia University's Hod Lipson remarked
after being asked to assess demands for more transparent AI systems.[51]
When it comes to killing cancer cells or predicting the weather, Lipson
may have a point; we do not need to understand the exact mechanism of
an AI in order to empower it to solve our problems. But when it comes to
important decisions about people, the inexplicable is inappropriate. As
the European Union's emerging "right to an explanation" shows, it can
be limited and replaced with a more humane approach.

Some of the biggest battles over robotics and AI will focus on the
analytic power of machines. What data are they allowed to gather and
use? And how will those data be processed? These questions are of vital
importance to the future of democracy and communications. Think
about how misinformation flourishes.[52] While biased propaganda has
long haunted the media, a largely automated public sphere has super-
charged it, enabling outright falsehoods and fabrications to spread virally. Some authorities have now begun to intervene, damping the spread
of hate speech and falsehoods. That is a first step toward repairing the
online public sphere, but much more will be required, including a more
prominent role for journalists who abide by traditional norms of their
profession.

Cyberlibertarians will argue that their AIs should enjoy a "freedom
to think" that includes the processing of any data they come across or

that their owners want to "feed" them. In the realm of pure computing unconnected to social consequences, that right might be respected. All manner of irresponsible speech is permitted in the name of free expression; software programmers can assert a similar right to enter data into programs without regard to its social consequences. But as soon as algorithms—and especially robotics—have effects in the world, they must be regulated and their programmers subject to ethical and legal responsibility for the harms they cause.[53]

PROFESSIONALISM AND EXPERTISE

Who gets to decide what this responsibility entails? A smooth and just transition will demand both old and new forms of professionalism in several key areas. The concept of expertise commonly connotes a mastery of a certain body of information, but its actual exercise involves much more.[54] For those who conflate occupational duties with mere knowledge, the future of employment looks grim. Computers' capacity to store and process information has expanded exponentially, and more data about what individuals do during their workday is constantly accumulating.[55] But professionalism involves something more complex: a recurrent need to deal with conflicts of values and duties, and even conflicting accounts of facts.[56] That makes a difference to the future of work.

For example, imagine that you are driving down a two-lane road at forty-five miles per hour, cruising home. You see a group of children walking home from school about a hundred yards ahead. Just as you're about to pass by them, an oncoming eighteen-wheeler swerves out of its lane and is about to hit you head on. You have seconds to decide: sacrifice yourself, or hit the children so you can avoid the truck.

I like to think that most people would choose the nobler option. As the automation of driving advances, such self-sacrificial values can be coded into vehicles.[57] Many cars already detect whether a toddler in a driveway is about to be run over by a driver with a blind spot. They even beep when other vehicles are in danger of being bumped. Transitioning from an alert system to a hardwired stop is technically possible.[58] And if that is possible, so is an automatic brake that would prevent a driver from swerving for self-preservation at the expense of many others.

But the decision can also be coded the other way—to put the car occupants' interests above all others. Although I do not think that that is the correct approach, the correctness of the approach is beside the point for our purposes. The labor question addresses how engineers, regulators, and marketers, as well as government relations and sales professionals, work together to shape human-computer interactions that respect the interests of everyone affected by automation, while also respecting commercial imperatives. There are few one-shot problems in design, marketing, and safety. As technology advances, users adapt, markets change, and new demands are constantly arising.

The medical profession has long been faced with such dilemmas. Doctors' jobs are never limited to merely taking care of the cases before them. Obliged to understand and monitor risks and opportunities that are constantly shifting, doctors must keep track of where medicine is headed, staying current about studies that either confirm or question mainstream medical knowledge. Consider even a decision as trivial as whether to give an antibiotic to a patient with a sinus infection. A good primary-care doctor must first decide whether the drug is clinically indicated. Doctors may take subtly different positions on how robust their obligation is to conserve antibiotic prescriptions to slow the evolution of resistant microbes. They also need to keep track of the prevalence of potential side effects of antibiotics, such as the sometimes devastating infections caused by *Clostridium difficile*—and the varying likelihood of such effects on different types of patients. Patients have some awareness of all these things when they visit a physician, but they are not responsible for coming to a correct decision or melding all these judgment calls into a recommendation for a particular case. That is the professional's role.

For true believers in the all-encompassing power of big data, predictive analytics, algorithms, and AI, the "brains" of robots can hack their way around all these problems. This is a tempting vision, promising exponential technological progress to raise living standards. But is it a realistic one? Even systems based purely in the digital realm—such as search-engine algorithms, high frequency trading, and targeted advertising—have proven in numerous cases to be biased, unfair, inaccurate, or inefficient.[59] Information is much harder to capture accurately in the wild, and there are disputes over what should be measured in the first place. Stakes rise considerably higher when algorithmic systems are empowered as the brains of robots that can sense their environment and act upon it. Meaningful human control is essential.

Nor is such human control only necessary in fields such as medicine, which has a long history of professional self-governance. Even in transport, professionals will have critical roles for decades to come. However fast robotic driving advances, the firms developing it cannot automate the social acceptance of delivery drones, sidewalk wagons, or cars. As legal expert Bryant Smith has observed, lawyers, marketers, civil engineers, and legislators must all help prepare society as a whole for the widespread deployment of such technologies.[60] Governments need to change their procurement policies, both for vehicles and infrastructure. Local communities must make difficult decisions about how to manage the transition, since stop lights and other road features optimized for human drivers may not work well for robotic vehicles, and vice versa. As Smith observes, "Long-term assumptions should be revisited for land-use plans, infrastructure projects, building codes, bonds, and budgets."[61]

The labor required for such a transition will be vast and diverse.[62] Security experts will model whether vehicles without human passengers pose special risks to critical infrastructure or to crowds. Terrorists do not need to recruit a suicide bomber if they can load a self-driving car with explosives. Public health experts will model the spread of infectious disease if such vehicles include strangers. Legislators are already grappling with the question of whether to require such vehicles to revert control to a person upon request or to give that control to police when they demand it.[63] I used the ambiguous term "person" in the last sentence because we still do not have a good term for occupants of a semi-autonomous vehicle. Both law and norms will shape that new identity over time.[64]

None of these decisions should be made solely—or even predominantly—by the programmers and corporations developing algorithms for self-driving cars. They involve governance and participation by a much wider range of experts, ranging from urban-studies scholars to regulators to police and attorneys. Negotiations among affected parties are likely to be protracted—but that is the price of a democratic and inclusive transition toward new and better technology. And these are only a few of the ethical, legal, and social implications of a widespread transition to self-driving cars.[65]

Nevertheless, some futurists argue that AI obviates the need for professions. With a large enough set of training data, they argue, virtually any human function can be replaced with a robot. This book takes the exact opposite view: to the extent our daily lives are shaped by AI and machine learning (often run by distant and massive corporations), we need more and better professionals. That is a matter of affirming and extending the patterns of education and licensure we already have in fields like medicine and law. And it may require building entirely new professional identities in other critical sectors where both wide public participation and expertise are essential.

TWO CRISES OF EXPERTISE

Asserting humans' value as a form of expertise may be jarring to some readers. At present, the most popular argument against AI's encroachment on the governance of workplaces and cities is a democratic appeal. AI's critics argue that technical experts in topics like machine learning and neural networks are not diverse enough to represent the persons their technology affects.[66] They are too removed from local communities. The same could be said of many other experts. There is a long and distinguished history of activists complaining about aloof doctors and professors, incomprehensible lawyers, and scientists detached from the quotidian problems of everyman. Confronted about economists' predictions of disastrous consequences from Brexit, British politician Michael Gove asserted that "people in this country have had enough of experts."[67] As that sentiment fuels populist campaigns around the world, there is a deepening chasm between politics and expertise, mass movements and bureaucratic acumen, popular will and elite reasoning.

Commenting on such trends, the sociologist Gil Eyal argues that expertise is a way of talking about the "intersection, articulation, and friction between science and technology on the one hand, and law and democratic politics on the other."[68] This is indeed a venerable tension in administration, where bureaucrats must often make difficult decisions implicating both facts and values. For example, raising or reducing pollution limits is a decision with medical consequences (for the incidence of lung cancer), economic impact (on the profitability of enterprise), and even cultural significance (for the viability of, say, mining communities). Eyal focuses on a democratic challenge to pure technocratic decision-making on each of those fronts.

This book examines a different, and distinct, challenge to expertise—or, more precisely, a clash of forms of expertise. Well-credentialed economists and AI experts have asserted that their ways of knowing and ordering the world should take priority almost everywhere, from hospitals to schools, central banks to war rooms. Few are as blunt as a former technology company CEO who remarked to a general, "You absolutely suck at machine learning. If I got under your tent for a day, I could solve most of your problems."[69] But the general theme of many books on AI-driven automation and economic disruption is that the methods of economics and computer science are *primus inter pares* among other forms of expertise. They predict (and help enact) a world where AI and robotics rapidly replace human labor, as economics dictates cheaper methods of getting jobs done. On this view, nearly all workers will eventually share the fate of elevator operators and horse-and-buggy drivers, just waiting until someone with adequate data, algorithms, and machinery replaces us.

To be sure, there are areas where economics and AI are essential. A business cannot run without covering its costs; a self-checkout lane will fail if its scanner program cannot recognize product bar codes. But the questions of whether a business should exist or a cashier should be replaced with a robot kiosk cannot be answered by economics or computer science alone. Politicians, communities, and businesses decide, based on complex sets of values and demands. These values and demands cannot simply be reduced to equations of efficiency and algorithms of optimization, entirely distinct and abstracted from communities. Rather, they are expressed and reconciled by human experts, either in current or future professions capable of demonstrating how their workers' and managers' intimate, local knowledge results in services and practices worth preserving.

To pursue the opposite—unleashing commercial pressures and robotic mimicry to colonize, replace, and rule every form of human labor—is to radically reorganize society. The sociologist Will Davies once observed that "a profession that claimed jurisdiction over *everything* would no longer be a profession, but a form of epistemological tyranny."[70] Today, far too many discussions of AI and robotics are dominated by a narrow focus on efficiency and optimization. My goal here is to bring in a far greater range of goals and values to the automation conversation. To entrench that richer dialogue, we need to ensure that wherever work and service expresses and reflects our values, *distributed expertise* melds the democratic value of representation with the cognitive values of accuracy, effectiveness, and scientific method.[71]

This approach is a far cry from that of many AI programmers in the 1980s who aspired to boil down the decision-making process of lawyers and doctors into a series of "if, then" decision trees. For example, such a program may ask the physician-trainee, "Does the patient have a fever? If so, then ask when the fever started. If not, ask if the patient has a cough." There was enormous enthusiasm for such systems, but they remained cumbersome and inconvenient. Professional judgment turned out to be much harder to systematize than AI researchers expected.

The philosopher Hubert L. Dreyfus developed theories of tacit knowledge to explain why expert systems performed so poorly.[72] We know more than we can explain. Think of how difficult it would be to boil down your job into a series of "if, then" statements. Could the situations you encounter daily be recognized by a computer? Would possible responses to those situations also be readily articulated, evaluated, and ranked? If not (and I suspect for the vast majority of readers, this is the case), the difficulties of description (of job tasks) and translation (of human situations into computable code) give an important insight into the enduring role of humans in work.

To be sure, variation may be something to be checked, not celebrated. If one set of otolaryngologists orders tonsillectomies for 90 percent of the children it treats, when all others have a rate of 15 percent or lower, there may well be serious problems of incompetence or opportunism to investigate.[73] Still, the autonomy of physicians to innovate or to vary from what the majority do is often protected in law and desired by patients.[74] There is too much uncertainty in ordinary medical practice to reduce it all to algorithms, which are commonly derided as "cookbook medicine." Patients also want to be seen by someone with some empathy for their plight and some direct personal encouragement that they improve. Similarly, most parents probably would not like to see their children taught by a single nationally broadcast teacher, however much that might reduce the taxes they pay for schools. There is a common sense that a direct connection to a trusted person is far more valuable than the mere broadcast of even the most skilled person—or robot.

Thus, one way to alleviate the democratic crisis of expertise—the tension between aloof technocrats and passionate populists—is to empower local professionals. We do not want classroom curricula dictated in detail by state legislatures or multinational corporations; instead, teachers and professors with some ongoing common sense of what is important in their fields need to have ample opportunity to enrich and enliven the basics. Expand a sense of the value of this "personal touch" in many other fields—its importance to workers and those served by them—and the basic case against premature automation emerges. If AI is to succeed, it must at the very least be fed diverse data streams from human observers. "Success" may even be defined as assistance to professionals knowledgeable and committed enough to know when to trust the machine and when to rely on their own judgment.

THE BENEFIT OF COST

Technocratic visions of rapid, expansive automation generate a strange tension at the core of contemporary economic policy. When the question of technological unemployment confronts, say, the US Council of Economic Advisors, the World Economic Forum, or the International

Monetary Fund, experts sternly warn that tens of millions of jobs are about to be replaced by robots.[75] Focused on our role as producers, this is a discussion framed by gloom and urgency. Field after field, it seems, is set to be automated—first routine tasks, then more professional roles, and then even the work of coding itself once a "master algorithm" has been found.[76] Reporting on this literature can be apocalyptic. "Robots to steal 15 million of your jobs," blasted the *Daily Mail,* trumpeting a study touted by Bank of England governor Mark Carney.[77] While estimates of job loss vary widely, the economic literature's drumbeat is unmistakable: every worker is at risk.

Simultaneously, economists celebrate the cheapening of services. The model of economic progress here is eerily similar to the one featured in automation narratives. Leaders in the health care and education sectors are supposed to learn from the successes of the assembly line in manufacturing, as well as data-driven personalization in the internet sector. Dialectically templatized and personalized approaches to health and education are supposed to make hospitals and schools cheaper and eventually make the best of their services available to all.[78]

Combine "robots are taking all the jobs" dystopianism with "ever-cheaper services" utopianism, and you have a bifurcated vision of our economic future. The workplace is destined to become a Darwinian hellscape, where employees are subordinate to machines that record their every movement to develop robotic replicants. The only consolation comes after hours, when the wonders of technology are supposed to make everything cheaper.

This model of miserable workers and exultant consumers is not merely troubling—it is unsustainable. Taken individually, a reduction in labor costs seems like a good thing. If I can replace my dermatologist with an app and my children's teachers with interactive toys, I have more money to spend on other things. The same goes for public services; a town with robot police officers or a nation with drone soldiers may pay less taxes to support their wages and health care. But doctors, teachers, soldiers, and police are all potential purchasers of what others have to sell. And the less money that they have, the less money I can charge them. In classical economic terms, the great worry here is deflation, a self-reinforcing spiral of lower wages and prices.

Even in the most self-interested frame, the "cost" of goods and services to me is not a pure drain on my well-being. Rather, it is a way of reallocating purchasing power to empower those who helped me (by creating whatever I am buying) to eventually help themselves (to perhaps purchase what I make or do). To be sure, a universal basic income would make up some of the purchasing power of those put out of work by robotics. But it is unrealistic to expect redistribution to do anything near the work of "pre-distribution" in assuring some balanced pattern of economic reward. Most democratic electorates have been cutting the relative tax liability of the richest for decades.[79] Robotization is unlikely to change that dynamic, which unravels ambitious plans for redistributing wealth.

Regarding the economy as an ongoing ecology of spending and saving, and as a way of parceling out power over (and responsibility for) important services, gives us a better perspective on the robotics revolution. Traditional cost-benefit analysis tends to dictate a rapid replacement of humans by machines, even when the machines' capabilities are substandard. The lower the cost of a service, the greater its benefits appear by comparison. But once we understand the benefits of cost itself—as an accounting of effort and an investment in persons—the shortcomings of this simplistic, dyadic view of the economy becomes clearer. The penultimate chapter further develops the benefits of the costs of programs and policies recommended in the rest of this book.

PLAN OF THE BOOK

Too many technologists aspire to rapidly replace human beings in areas where we lack the data and algorithms to do the job well. Meanwhile, politicians have tended toward fatalism, routinely lamenting that regulators and courts cannot keep up with technological advance. The book will dispute both triumphalism in the tech community and minimalism among policymakers in order to reshape public understanding of the state's role in cultivating technological advance. I offer policy analysis that shows the power of narrative and qualitative judgment to guide the development of technology now dominated by algorithmic methods and quantitative metrics. The point of the book is to distill accumulated knowledge from many fields and vantage points and present it to the public for its use. Ideally, it will lay a foundation for what Alondra Nelson calls "anticipatory social research," designed to shape, and not merely respond to, technological advance.[80]

The translation of tasks into code is not a purely technical endeavor. Rather, it is an invitation to articulate what really matters in the process of education, caregiving, mental-health care, journalism, and numerous other fields. Although there is a temptation to simply set forth quantifiable metrics of success in all those fields and to optimize algorithms to meet them (whether via trial and error, crunching past data, or other strategies), the definition of what counts as success or failure in such fields is highly contestable. Basing a decision on one metric excludes all others. No one can be "driven" by data in general. Particular data matter, and the choice of what to count (and what to dismiss as non-representative) is political.

Among AI ethicists, tension has developed between pragmatists (who focus on small and manageable reforms to computational systems to reduce their discriminatory or otherwise unfair judgments) and futurists, who worry about the rise of out-of-control systems and self-improving AI (which could, it is feared, rapidly grow "smarter," or at least more lethal, than their human creators). The pragmatists tend to dismiss the futurists as haunted by phantoms; the futurists think of the pragmatists' concerns as small bore. I believe each side needs the other. The horrific outcomes predicted by futurists are all the more likely if we do not aggressively intervene now to promote transparency and accountability in automated systems. But we are unlikely to take on that hard task if we fail to reckon with the fundamental questions about human nature and freedom that the futurists are asking.

These questions are not new. For example, in 1976, computer scientist Joseph Weizenbaum asked, "What human objectives and purposes may not be appropriately delegated to computers? . . . The question is not whether such a thing can be done, but whether it is appropriate to delegate this hitherto human function to a machine."[81] But the queries

"Can robots be better than humans?" or "When should humans not use robotics?" are incomplete. Almost anyone, in any job, is already using some level of automation on a continuum between simple tool and human-replacing AI. A better frame is, "What sociotechnical mix of humans and robotics best promotes social and individual goals and values?"

In a series of case studies, I answer that question concretely by making the case that AI supplementing, rather than replacing, human expertise realizes important human values. Chapters 2, 3, and 4 describe what that process might look like in health care, education, and media, focusing on the first new law of robotics: the need for technology to complement, rather than replace, existing professionals.

I am, on the whole, optimistic about the prospects for complementary automation in health and education. Patients and students by and large demand human interaction.[82] They realize that however advanced AI becomes, it is enormously helpful to obtain guidance on how to use it from experts who study the reliability of various sources of knowledge daily. Even more importantly, in so many care or learning contexts, human relations are intrinsic to the encounter. Robotic systems can provide technical support, improving judgments and developing entertaining and engaging drills. Perhaps rural and disadvantaged areas will demand them as substitutes for now-absent professionals. But this dictate of necessity is far from an exemplary labor policy. It is particularly troubling when it comes to mental health care for vulnerable populations.

When there is a nurse, teacher, or doctor at the point of contact of AI—to mediate its effects, assure good data collection, report errors, and do other vital work—there is far less chance of a grimly deterministic future in which we are all poked and prodded into learning or wellness by impersonal machines. Professionals in health and education also owe clear and well-established legal and ethical duties to patients and students. These standards are only beginning to emerge among technologists. Thus, in the case of media and journalism—the focus of Chapter 4—a concerted corrective effort will be necessary to compensate for what is now a largely automated public sphere.

When it comes to advertising and recommendation systems—the lifeblood of new media—AI's advance has been rapid. Reorganizing commercial and political life, firms like Facebook and Google have deployed AI to make the types of decisions made by managers at television networks or editors at newspapers—but with much more powerful effects. The reading and viewing habits of hundreds of millions of people have been altered by such companies. Disruption has hit newspapers and journalists hard. And it has been terrifying for some vulnerable groups, including minorities targeted for harassment. The only way to stem an epidemic of fake news, digital hate campaigns, and similar detritus is to bring more responsible persons back in to guide the circulation of online media.

Whereas Chapter 4 focuses on AI's failures in judging the value of news, Chapter 5 describes the perils of using AI to judge people. Computation is playing an ever larger role in hiring and firing, as well as in the allocation of credit and the treatment of debt. It is also creeping into security services. I warn against the rapid adoption of robotic policemen and security guards. Even predictive policing, a wholly software-driven affair supervised by officers, has proven controversial thanks to its frequent reliance on old and biased data. Machines sent out on their own to patrol neighborhoods or rustle panhandlers off sidewalks are even more troubling. Nor are many other applications of AI in civil law ready for widespread adoption. They promote a rule of machines over persons, which sacrifices human dignity on the altar of efficiency.

Chapter 6 takes this concern about power to traditional and online battlefields. Debate on lethal autonomous weapon systems has taken on a familiar structure. Abolitionists call for a ban on killer robots, realists reject that approach, and reformers occupy a middle ground by proposing regulation short of an outright ban. Abolitionists and reformers are now engaged in spirited arguments about the value of each side's approach. But the two groups' strategies may eventually harmonize. Reformers acknowledge that some types of weapons systems are so dangerous that they should never be built. Abolitionists concede that some defensive uses of automation (particularly in cyberwarfare) are necessary for national security.

Both abolitionists and regulators deserve great credit for engaging in lawfare to impose limits on the tactics available in conflict. But their hard work could be unraveled by arms race dynamics that realists both recognize and promote. The ubiquity of LAWS matters, and hinges on the degree to which leading powers invest in cyberwarfare, robotic soldiers, and other AI-enhanced weaponry. Retarding such investment—a key corollary of the third new law of robotics I described above—will require citizens and leaders to accept limits on states' power to project force.

The reforms I propose will be expensive. But that expense can be seen as much as a solution to the problems posed by the robotization of force as it is a problem for the public fisc. Chapter 6 argues that governments forced to escalate the quality of (and access to) robust health and education sectors are less likely to have the resources to automate violence than those unencumbered by such obligations. At bottom, a new and better global political economy is the only way to avoid a tragic drain of resources into a vortex of robotic arms races of surveillance and countersurveillance, force and deflection of force.

Chapter 7 maps the basic features of a renewed political economy of automation, including fiscal and monetary policy. Each must adjust in order to promote human-centered AI. Universal basic income proposals are, at present, in the spotlight. As the limits of redistributive policy become clearer, guarantees of universal basic services and jobs will be even more critical. These policies aim not only at guaranteeing subsistence, but also ensuring more democratic governance of the economy as a whole. They correct for the many other policies prioritizing capital accumulation over labor, and substituting machines for persons.

Admittedly, neither substitutive nor complementary conceptions of robotics are irrefutably entailed from data. Rather, they hinge on larger visions of the nature and purpose of human work and technology. Those visions entail a certain way of interpreting not just economic reality, but also the culture of robotics and employment. The last chapter explores these cultural questions, including the stories we tend to tell about AI and robotics. These narratives are, at their best, not merely symptoms of anxiety or aspiration, but sources of wisdom.

We can uphold a culture of maintenance over disruption, of complementing human beings rather than replacing them. We can attain and afford a world ruled by persons, not machines. The future of robotics can be inclusive and democratic, reflecting the efforts and hopes of all citizens. And the new laws of robotics can guide us on this journey.

2

▪ Healing Humans

There are two AI dreams in medicine. The first is utopian, straight out of science fiction novels. Care robots will spot and treat any disease, instantly. Nanobots will patrol our veins and arteries, busting clots and repairing damaged tissues. Three-dimensional printed organs, bone, and skin will keep us all looking and feeling young well into our eighties and nineties. With enough luck, even brains can be uploaded for perpetual safekeeping, with robotic bodies sleeving indestructible minds.[1]

Whatever its long-term merits, that sci-fi vision is far, far off—if it ever arrives at all. More realistic medical futurists are still ambitious, but they offer more attainable visions. They recognize the critical role that human empathy plays in care, that human insight contributes to diagnoses, and that human dexterity adds to surgery. They by and large embrace the first new law of robotics, promoting a future where AI is primarily aiding (rather than replacing) doctors and nurses. That is wise, reflecting a realistic assessment of the current state of technology and data.[2]

Unfortunately, even many realists tend to stumble when it comes to policy and law. They see health care systems through a primarily economic lens, lamenting their expense and inefficiencies. They advocate deregulation to spur innovation and budget limits to force cost cutting. But what we really need in medical technology policy is more responsibility to collect and use the best data—not less. And we need to invest in the cutting edge of medical practice, rather than simply assuming that hospitals and doctors will come up with ever more ingenious ways of doing more with less.[3]

Science fiction writers dream of a day when a combination of apps and robots can take care of all our medical needs. But this is not the current path of leading medical technological developments—nor should policymakers intervene to make it our lodestar. The stakes of medical care, as well as the psychological stress of illness, counsel in favor of a lasting human presence in the deployment of health-sector automation. While economic imperatives will pressure hospitals and insurers to substitute software for therapists and bots for nurses' attention, professional associations should ensure that cost considerations are balanced against the many virtues of direct human involvement in care.

DECIDING WHEN TO SEEK HEALTH CARE

Imagine waking up with a piercing stomach pain. Is it appendicitis? Bloating? A pulled muscle? Stomach pain is one of the hardest differential diagnoses for even seasoned emergency doctors. Abdominal agony could result from any one of dozens of conditions, ranging from the trivial to the life threatening.[4] Even a slight risk of a disastrous outcome would seem to counsel in favor of a trip to the hospital to get some professional advice—stat.

For the wealthy or well insured, the decision can be an easy one. For others, though, it could be ruinous to seek help. In the developing world, medical bills may threaten families' ability to meet basic needs.[5] In the United States, millions are either uninsured or underinsured. A single trip to the ER can cost well over $10,000; even a false alarm can run into the thousands of dollars once tests, physician fees, and other charges are taken into account. Even for those with adequate insurance and ample financial resources, a hospital visit poses the risk of unnecessary tests, exposure to viruses, and hours of inconvenience.

For many, the first place to look for information about sudden symptoms is Google. And for many years, Google saw medical searches—like those prompted by a sudden pain in the middle of the night—as little different than other searches. As long as they had enough "Google juice" (that mysterious mix of relevance and authority that powers content to the top of search results), sites of dubious reliability might mix with information from established doctors or medical schools. The burden was on Google users to separate the wheat from the chaff, discerning the credibility of sites.

By 2016, the company had revised its approach.[6] It collaborated with experts at the Mayo Clinic to vet information that appeared in common health searches.[7] Type in "headache on one side," and above or next to a standard list of search results, there appear a series of boxes, each briefly describing a possible classification for the headache. Pick any one of them (say, "tension headache"), and you find another box, again attributable to Google itself. It describes in very brief terms whether the condition is common, how common it is in various age groups, and what types of medical intervention might be helpful.

These new Google results are a heartening sign for artificial intelligence in health care. They do not reflect a company dead set on replacing the expertise of doctors with big data and algorithms. Rather, professionals are invited in, to help devise structured approaches to health care information and the health system itself. Similarly, IBM has shifted the marketing of its Watson system in health care and law, billing it as more of a helper for than a replacement of doctors.[8] When I talked to a representative of IBM's Watson team in 2017, he said that they promote a vision of augmented, not artificial, intelligence. As even the firms with the most to gain from AI marketing shift toward an IA (intelligence augmentation) approach, the dream of a wholly automated diagnostic tool may soon seem more anachronistic than futuristic. There will always be a place for domain experts to evaluate the accuracy of AI advice, and to assess how well it works in the real world.

AI'S CORE COMPETENCE: AVOIDING COMMON ERRORS

Doctors are expert pattern recognizers. We expect a dermatologist to tell us whether a mole is malignant or merely a beauty mark; we suffer colonoscopies to give gastroenterologists a chance to spot (and excise) polyps. Yet even the best doctor can make a mistake, and average physicians may become bored or distracted at critical moments. Thanks to AI, we may greatly reduce those types of errors, saving thousands of lives per year.

The method depends on massive amounts of data. A database may include labeled images of millions of different abnormalities that eventually became cancerous, as well as millions that did not. As we might search on Google for websites matching a query, a computer can rapidly compare images of your colon or skin with those in the database. Ideally, machines learn to spot "evil digital twins"—tissue that proved in the past to be dangerous, which is menacingly similar to your own.[9]

This machine vision—spotting danger where even experienced specialists might miss it—is far different from our own sense of sight. To understand machine learning—which will come up repeatedly in this book—it is helpful to compare contemporary computer vision to its prior successes in facial or number recognition. When a facial recognition program successfully identifies a picture as an image of a given person, it is matching patterns in the image to those in a preexisting database, perhaps on a 1,000-by-1,000-pixel grid. Each box in the grid can be identified as either skin or not skin, smooth or not smooth, along hundreds or even thousands of binaries, many of which would never be noticeable by the human eye. And even more sensing is available. Medical images may also encode data at the pixel or voxel (3-D pixel) level that map to what our hands, nose, or ear might sense—and far more.

Pattern recognition via machine vision found early commercial success with banks, which needed a way to recognize numbers on checks (given the wide variety of human handwriting). With enough examples of written numbers and computational power, this recognition can become nearly perfect. Thus, machine vision is "superhuman" in many respects, in terms of both "ingesting" data and comparing those data to millions of other images. A dermatologist might use a heuristic to diagnose melanoma (such as ABCDE, for asymmetry, borders that are irregular, color that is varied, large diameter, and evolving), or his or her experience of past cancerous and non-cancerous moles. A sufficiently advanced AI can check any of those ABCDE parameters against other moles with exceptional precision—so long as the data are accurate. Moreover, as sensors grow more advanced, AI may find unexpected sources of insight on what distinguishes malignant from benign moles.

Machine vision has "subhuman" aspects as well, and can exhibit surprising fragility.[10] Most of its applications in medicine now are "narrow AI," focused on a particular task and that task alone. Narrow AI for detecting polyps, for example, might "see" a problem polyp that no gastroenterologist would, but it might also be incapable of recognizing other abnormalities that it was not trained to detect. Joint work in diagnostics—involving both an AI program and a doctor—is more valuable than either working alone.[11]

Physicians train for years, but medical knowledge never stops advancing. It is no longer humanly possible to memorize every potential interaction between drugs—particularly in complex cases, where patients may be taking twenty or more medications. Pharmacists can play a role in stopping bad outcomes, but they, too, can overlook unusual problems.[12] Integrated into electronic health records, clinical decision support software (CDSS) is an early form of AI that can help physicians avoid terrible outcomes.[13] CDSS "monitors and alerts clinicians of patient conditions, prescriptions, and treatment to provide evidence-based clinical suggestions."[14] There is already evidence that CDSS reduces errors.[15] Yet even in this relatively straightforward area of information provision, programmers, managers, and engineers did not simply impose CDSS onto medical practice. Law has played a major role in its diffusion, including government subsidies to support such systems. The threat of malpractice lawsuits (for doctors) or corporate liability (for hospitals) counsels in favor of adopting CDSS; however, courts have also recognized that professional judgment cannot be automated, and they are loath to make failure to follow a machine recommendation an automatic trigger for liability if there are sound reasons for overriding the CDSS.[16]

Ongoing regulation will be critical here to assure that patients will have the benefit of cutting-edge technology, without burdening doctors and nurses with information overload. Several authors have chronicled the problem of "alert fatigue."[17] Human-computer interaction experts are working to assure a better balance between alerts and subtler reports about potential problems. The ideal CDSS software should be neither overbearing nor merely a quiescent watcher of practitioners. It can only fulfill its promise if its messages are continually calibrated and recalibrated to ensure that physicians, nurses, and pharmacists actually welcome its use, and have ongoing opportunities to critique and improve it.

DATA, DISCRIMINATION, AND HEALTH DISPARITIES

Human-validated AI may well become the standard of care in medicine, what every patient is due when seeking treatment.[18] Yet the mere fact that a technology is better in general does not mean it is optimal for all cases.[19] Indeed, AI can easily end up discriminating when it is based on flawed data. A well-known "risk score" (used to target assistance to patients in particular need) prioritized white patients over black patients because it used cost of care as a proxy for severity of illness. African Americans tended to receive less expensive care, so they were judged by the algorithm as less in need of medical attention than they actually were. The dangers of such proxy measures should be well-known, but in the rush to quantify, risks are often ignored or downplayed.[20]

Bias may also infect diagnostic AI. Physicians and computer scientists are already concerned that melanoma–detecting software will fail to work as well for minorities, who tend to be less well represented in training data sets.[21] If this disparity does indeed play out, one question for the legal system is whether the standard for detecting diseased skin in general will rise to the level achieved by such systems among those of the dominant ethnicity. Regulators need to ensure that more representative data is readily available, and that it is used. Otherwise, they risk further entrenching already alarming health disparities.

If regulators fail to step up, courts will need to decide when certain additional data is necessary to a given project (like identifying melanomas), and when it is simply "something nice to have" for those able to purchase high-end care. Malpractice law is designed to give patients reassurance that if their physician falls below a standard of care, a penalty will be imposed and some portion of it dedicated to their recovery.[22] If providers fail to use sufficiently representative datasets to develop their medical AI, lawsuits should help hold them accountable, to ensure that everyone benefits from AI in medicine (not just those lucky enough to belong to the most studied groups).

Data scientists sometimes joke that AI is simply a better-marketed form of statistics. Certainly narrow AI, designed to make specific predictions, is based on quantifying probability.[23] It is but one of many steps taken over the past two decades to modernize medicine with a more extensive evidence base.[24] Medical researchers have seized on predictive analytics, big data, artificial intelligence, machine learning, and deep learning as master metaphors for optimizing system performance. Literature in each of these areas can help regulators identify problematic data in AI. Moreover, critiques of the limits of AI itself (including lack of reproducibility, narrow validity, overblown claims, and opaque data) should also inform legal standards.[25] The key idea here is that AI's core competence—helping humans avoid errors—must now be turned on the humans who create AI. They must be held responsible for failures to use proper data and methods. Otherwise, we could just end up replicating errors in a future that should be free of them.

Activists are now exposing numerous examples of problematic datasets in medicine. For example, Caroline Criado Perez has proven that in far too much medical research and pedagogy, maleness is assumed as a default.[26] As she observes, "Women are not just smaller men: male and female bodies differ down to a cellular level . . . [but] there are still vast medical gender data gaps to be filled in."[27] The biases of datasets become even more inexcusable as work like Criado Perez's makes them more widely known. We need to fund much better data collection to ensure fair and inclusive AI in medicine—and to impose duties on developers, doctors, and hospitals to use it.

FOUR HORSEMEN OF IRRESPONSIBILITY

Thanks to long-standing campaigns to limit liability, there will be widespread resistance to such duties. AI also poses new barriers to accountability—and not just in medicine. Futurists envision AIs that effectively act of their own accord, without direction or control by their developers (or any other person). How, the challenge goes, can the creators or owners of such general-purpose technology anticipate all the potential legal problems that their AI might generate or encounter? No one wants to hold Microsoft responsible for ransom notes written as an MS Word document, which is a blank slate. Nor are parents responsible for the crimes of their adult children, who are independent entities.

When they assert that they are not responsible for their creations, leading developers of AI benefit from both the "blank slate" and "independent entity" metaphors. Given a decade of research on algorithmic accountability, neither justification should immunize such firms. We all now know that algorithms can harm people.[28] Moreover, lawyers have grappled with the problem of malfunctioning computers for decades, dating back at least to the autopilot crashes of the 1950s and the Therac-25 debacle of the 1980s (when a software malfunction caused tragic overdoses of radiation).[29]

Nevertheless, some proposals would severely diminish the role of courts in the AI field, preempting their traditional role in assigning blame for negligent conduct. Others would kneecap federal regulatory agencies, leaving it up to judges to determine remedies appropriate for accidents. Even if such legal "reforms" never happen, firms could limit or shift their liability via infamous terms of service that consumers "agree to" as contracts. Finally, free expression absolutists argue that when AI is simply "saying" things about persons rather than doing things to them, it should be treated like free speech and be immune from lawsuits. Advocates for these four horsemen of irresponsibility—sweeping preemption, radical deregulation, broad exculpatory clauses, and opportunistic free expression defenses—argue that AI will develop rapidly only if inventors and investors are free from the threat of lawsuits.

Bewitched by the promise of innovation, policymakers may be tempted to sweep away local laws in order to give industry leaders an immediate, very clear picture of their legal obligations.[30] Or they may "empower" AI users to contract away their rights to sue. The perverse case for contractual sovereignty here is that my right to give away my rights advances my autonomy. A less strained, utilitarian rationale is that is that citizens need to give up some rights so that AI can flourish.

Even if liability shields are needed to spur some innovation, they must not be absolute. As Wendy Wagner has observed, tort litigation is critical to exposing information that may be blocked from regulators.[31] When regulation is harmonized internationally or for a nation, more local entities should also be empowered to develop their own standards for the level of risk they are willing to accept from new technology.[32] While that granular litigation and regulation moves forward, higher-level authorities have the resources and time frame to map out broad trends in technological development and to solicit expert advice. For example, the US National Committee on Vital and Health Statistics (where I began a four-year term in 2019) offers policymakers expert advice on how data are optimally collected, analyzed, and used. That advice is critical because in well-ordered societies, regulators help shape technology's development (and do not merely react to it once it is created).[33]

Moreover, courts and legislatures should be wary of exculpatory clauses, limiting when consumers can sign away their rights. Judges have frequently been unwilling to recognize such clauses in the medical context, reasoning that patients are vulnerable and lack the information necessary for a truly informed choice.[34] We all stand in a similar position of vulnerability when it comes to most robotics and AI, since we are almost never privy to the data and codes behind them. Even where exculpatory clauses are allowed, there is still an important role for courts to play in policing unfair terms.[35] And there are certain types of causes of action that should be preserved, whatever terms contracting parties are willing to agree to.

In order to assess risks responsibly, both vendors and users need accurate accounts of the data used in AI (inputs) and data on its performance (outputs). No one should be allowed to contract away the right to inspect those data when AI causes harm.[36] The next section describes how regulators can help assure better inputs of data to AI-driven innovation and how they can promote quality outputs from such technology.

WHO WILL TEACH THE LEARNING HEALTH CARE SYSTEM?

We might once have categorized a melanoma simply as a type of skin cancer. But that is beginning to seem as outdated as calling pneumonia, bronchitis, and hay fever "cough." Personalized medicine will help more oncologists gain a more sophisticated understanding of a given cancer as, say, one of a number of mutations. If they are properly combined, compared, and analyzed, digitized records could indicate which combination of chemotherapy, radioimmunotherapy, surgery, and radiation has the best results for that particular subtype of cancer. That is the aspiration at the core of "learning health care systems," which are designed to optimize medical interventions by comparing the results of natural variations in treatments.[37]

For those who dream of a "Super Watson" moving from conquering *Jeopardy* to running hospitals, each of these advances may seem like steps toward cookbook medicine implemented by machine. And who knows what's in the offing a hundred years hence? In our lifetime, what matters is how all these data streams are integrated, how much effort is put into that aim, how participants are treated, and who has access to the results. These are all difficult questions, but no one should doubt that juggling all the data will take skilled and careful human intervention—and plenty of good legal advice, given complex rules on health privacy and human subjects research.[38]

To dig a bit deeper in radiology: the imaging of bodily tissue is rapidly advancing. We've seen the advances from X-rays and ultrasound to nuclear imaging and radiomics.[39] Scientists and engineers are developing ever more ways of reporting what is happening inside the body. There are already ingestible pill-cams; imagine much smaller, injectable versions of the same.[40] The resulting data streams are far richer than what came before. Integrating them into a judgment about how to tweak or entirely change patterns of treatment will take creative, un-systematizable thought. As radiologist James Thrall has argued,

> The data in our . . . information system databases are "dumb" data. [They are] typically accessed one image or one fact at a time, and it is left to the individual user to integrate the data and extract conceptual or operational value from them. The focus of the next 20 years will be turning dumb data from large and disparate data sources into knowledge and also using the ability to rapidly mobilize and analyze data to improve the efficiency of our work processes.[41]

Richer results from the lab, new and better forms of imaging, genetic analysis, and other sources will need to be integrated into a coherent picture of a patient's state of illness. In Simon Head's thoughtful distinction, optimizing medical responses to the new volumes and varieties of data will be a matter of *practice,* not predetermined *process.*[42] Both diagnostic and interventional radiologists will need to take up difficult cases anew, not as simple sorting exercises.

Given all the data streams now available, one might assume that rational health policy would deepen and expand the professional training of radiologists. But it appears that the field is instead moving toward commoditization in the US.[43] Ironically, radiologists themselves have a good deal of responsibility here; to avoid nightshifts, they started contracting with remote "nighthawk" services to review images.[44] That, in turn, has led to "dayhawking" and to pressure on cost-conscious health systems to find the cheapest radiological expertise available—even if optimal medical practice would recommend closer consultations between radiologists and other members of the care team for both clinical and research purposes. Government reimbursement policies have also failed to do enough to promote advances in radiological AI.[45]

Many judgment calls need to be made by imaging specialists encountering new data streams. Presently, robust private and social insurance covers widespread access to radiologists who can attempt to take on these challenges. But can we imagine a world in which people are lured into cheaper insurance plans to get "last year's medicine at last year's prices"? Absolutely. Just as we can imagine that the second tier (or third or fourth or fifth tier) of medical care will probably be the first to include purely automated diagnoses.

Those in the top tier may be happy to see the resulting decline in health care costs overall; they are often the ones on the hook for the taxes necessary to cover the uninsured. But no patient is an island in the learning health care system. Just as ever-cheaper modes of drug production have left the United States with persistent shortages of sterile injectables, excluding a substantial portion of the population from high-tech care will make it harder for those *with* access to such care to understand whether it's worth trying.[46] A learning health system can make extraordinary discoveries, if a comprehensive dataset can fuel observational research into state-of-the-art clinical innovations. The less people have access to such innovations, the less opportunities we have to learn how well they work and how they can be improved. Tiering may solve medicine's cost crisis at present, but sets back future medical advances for everyone. Thus, there is a high road to advances in medical AI, emphasizing better access for everyone to improving quality of care, and a cost-cutting low road, which focuses on merely replicating what we have. Doctors, hospital managers, and investors will implement the high road, the low road, or some middle path. Their decisions, in turn, are shaped by a shifting health law and policy landscape.

For example, consider the tensions between tradition and innovation in malpractice law. When something goes wrong, doctors are judged based on a standard of care that largely refers to what other doctors are doing at the time. Malpractice concerns thus scare some doctors into conformity and traditionalism. On the other hand, the threat of litigation can also speed the transition to clearly better practice. No doctor today could get away with simply palpating a large tumor to diagnose whether it is malignant or benign. Samples usually must be taken, pathologists consulted, and expert tissue analysis completed. If AI methods of diagnosis become sufficiently advanced, it will be malpractice not to use them, too.

On the other hand, advanced automation may never get any traction if third-party payers, whether government or insurers, refuse to pay for it. Insurers often try to limit the range of care that their plans cover. Patients' rights groups fight for mandated benefits. Budget cutters resist, and when they succeed, health systems may have no choice but to reject expensive new technology.

Other regulatory schemes also matter. Medical boards determine the minimal acceptable practice level for doctors. In the United States, the Centers for Medicare and Medicaid Services help set the terms for graduate medical education via subsidies. Well funded, they can design collaborations with bioengineers, computer scientists, and statisticians. Poorly funded, they will go on churning out too many physicians ignorant of the statistical knowledge necessary to do their current jobs well, let alone critically evaluate new AI-driven technologies.[47]

The law is not merely one more set of hurdles to be navigated before engineers can be liberated to cure humanity's ills. The key reason health care employment has actually grown as a sector for the past decade are the legal mandates giving wide swaths of the population guaranteed purchasing power, whatever their wages or wealth. At their best, those legal mandates also guide the development of a health care system toward continuous innovation and improvement.

WHO IS YOUR THERAPY APP WORKING FOR?

Despite the challenges facing AI-enhanced diagnoses of cancers and other bodily ailments, we can be confident that technology will have a large and growing role in these areas. Doctors and researchers can agree on optimal performance standards and gradually improve machines to recognize and accomplish more. Matters become more complicated when we turn from physical to mental health, necessitating even more human involvement.

Like other vulnerable populations, those with mental illness are often referred to automated systems. Myriad programs in Apple's App Store and Google's Android Market claim to improve mental health. There are apps for depression, anxiety, addiction, and phobias designed to offer some soothing or encouraging digital interface—and sometimes far more. When prescribed by a therapist or other professional, these apps are promising. For instance, those seeking treatment for substance abuse can log their cravings, correlate relapses to particular triggers, and receive prompts or advice. However, when apps become more autonomous, unmoored from experts, serious problems can arise.

Sometimes apps fail to perform basic therapeutic duties. Journalists discovered that one AI therapist failed to alert authorities when a test subject wrote that she was a child suffering sexual abuse at home.[48] Qualified domain experts know about mandatory reporting laws, while technologists trained to "move fast and break things" may have no idea they exist. Unvetted and even dangerous apps proliferate, risking the health of those in need of more competent care.[49] Data-sharing policies of such apps are often incomplete or one sided, risking breaches of confidentiality that a therapist is duty-bound to keep.[50] The US Food and Drug Administration (FDA) has shirked its duties in this space, barely regulating such apps after repeated warnings from members of Congress that agency funding might be slashed if they interfered with innovation. Nor do other jurisdictions appear to be particularly aggressive at ensuring safety and effectiveness.[51]

Economic pressures fuel that laxity. Cheap mental health apps are a godsend for health systems pressed by austerity to cut costs, like Britain's National Health Service (NHS). Via an "NHS Apps Library," UK health authorities recommended apps for those suffering from depression and anxiety. Unfortunately, according to a study in the journal *Evidence Based Mental Health,* the true clinical value of most apps was "impossible to determine."[52] Only two of the apps studied applied validated metrics. It is also by no means clear that digital substitutes for extant mental health professionals will live up to the type of fiduciary and other standards that are expected of human providers.[53] Already raising worries in the realm of digital assistants, these trust concerns will only be more pronounced in health care.[54]

There is a history here, too. Quick-fix technological solutions can easily take in the unwary. Dangerous patent medicines lured the credulous for decades, with sometimes disastrous results.[55] These chemicals were particularly tempting for those unable to get to a doctor, just as some mental health apps are advertised as "better than nothing" for those who cannot access a therapist. More recently, Congress removed dietary supplements from full FDA regulatory authority, creating a Wild West of dubious "natural cures."[56] The efficacy of many such substances is doubtful. It's not surprising to see deregulation similarly deform digital diagnoses and interventions.

The long history of therapy programs offers some perspective on the current, crowded market for AI counseling. The MIT computer scientist Joseph Weizenbaum developed one of the most notable programs, ELIZA (named after the Eliza Doolittle of *Pygmalion,* since the program copied speech patterns of users). This early chatbot mimicked a Rogerian psychotherapist. In a 1966 paper, Weizenbaum offered the following as an example of the type of exchange that might occur:

> User: Men are all alike.
> ELIZA: IN WHAT WAY?
> U: They're always bugging us about something or other.
> E: CAN YOU THINK OF A SPECIFIC EXAMPLE?
> U: Well, my boyfriend made me come here.
> E: YOUR BOYFRIEND MADE YOU COME HERE?
> U: He says I'm depressed much of the time.
> E: I AM SORRY TO HEAR YOU ARE DEPRESSED.

At first, Weizenbaum was proud of his creation. He showed it off to both faculty and staff at MIT. Users readily "confided" in the software— so readily, in fact, that Weizenbaum soon terminated his work on it, worried that it could usurp other, richer forms of therapeutic encounter.[57] Given contemporary health technocrats' obsession with cutting costs, that concern was prophetic. In an era of "value-based purchasing," a $2.99 app may be an irresistible replacement for a psychiatrist.

App boosters may claim that it is unfair to bring up ELIZA when so much more advanced technology has developed over the past half century. Yet leading consumer-facing mental health apps tend to use similar text-based interfaces. The very popular Woebot app is designed to automate aspects of cognitive behavioral therapy. Woebot starts user conversations "on rails," tracking them by offering simple, dichotomous responses. For example, near the start of its interaction, Woebot may text, "My data from my study at Stanford shows that it takes about [14 days] for people to get the hang of [chatting with me] and start feeling better." The user cannot type in critical questions like, "What study from Stanford? How many participants were in it? What was the control group exposed to?" Rather, the two allowed responses I saw when testing the app were "Makes sense" or "Hmm . . ." Such binary responses come up repeatedly in the app, disciplining user response.

For someone conditioned by multiple-choice tests, simple phone games like *Candy Crush* and the like / not-like binaries of Facebook, Instagram, and Twitter, Woebot's response buttons may be a welcome continuation of frictionless digital drift. They take away the burden of explaining oneself, articulating a clever response, and deciding whether to be warm or skeptical. That may be a relief for those already burdened by depression.

By contrast, there is a treatment method—colloquially known as talk therapy—that deems such articulacy to be the very *point* of counseling, not something to be avoided.[58] The rise of therapy apps risks further marginalizing talk therapy in favor of more behavioristic approaches, which try to simply end, mute, overwhelm, or contradict negative thoughts, rather than exploring their context and ultimate origins. When mental health apps are driven by narrow AI, they both benefit from and reinforce a utilitarian approach to patients, which frames their problems simply as impediments to productivity.[59]

True, this concern also arose decades ago with the rise of psychopharmaceuticals.[60] Critics argued that patients were better off exploring the larger context of their problems than trying to medicate them away.[61] But at least in the case of drugs, there is some expert guidance required before a course of treatment begins. All too many therapy apps skip even this safeguard, directly marketing to would-be patients.

Another line of criticism focuses on users' vulnerability to undue manipulation by unknown forces.[62] Consider, for instance, a therapy app "treating" a worker who complains a great deal about his or her time on the job. The worker feels underpaid and undervalued and expresses those concerns to the app. There are diverse potential responses to such a problem. For example, the app might counsel assertiveness, pushing the worker to ask for a raise. At the other extreme, the app might prescribe a contemplative resignation to one's fate, urging appreciation of all one already has. Or it might maintain a studied neutrality, digging ever deeper into the reasons for the worker's unease. Guess which response an employer may want to see in wellness apps provided to its workers. Professions serve as a buffer, to short-circuit such conflicts of interest. They can serve as more neutral interpreters of the maladies brought before them than apps or AI dictated by third party payers, be they employers, insurers, or governments.[63]

Predictive analytics works best when there are clear and well-defined metrics for success. But in the case of many mental health concerns, there are multiple ways to define problems and their resolution. Different commercial models can encourage different ways of defining mental illness or its treatment. A free app with an advertising-based model may want to encourage users to return as often as possible. A subscription-based service would not necessarily optimize for "time on machine," but it might aim to use other forms of manipulation to promote itself. Self-reports of well-being may be an uncontroversial ground truth against which to measure the value of apps.[64] But the concept of wellness itself is being colonized by corporations and governments and tied intimately to more objective measures, like productivity.[65] What is lost in this colonization is the sense, as Martha Nussbaum so eloquently evokes, of signifying emotions as "forms of judgment."[66] Instead, they risk becoming one more object of maximization via the digital soma of a burgeoning "happiness industry."[67]

Both academics and activists have pointed out problematic biases among doctors and other providers.[68] An algorithmic accountability movement in medicine will need to carry this work forward in critiquing—and fixing—the biases and other problems that will afflict computationally inflected caregiving. The best structural safeguard is to assure that most apps are developed as intelligence augmentation for responsible professionals rather than as AI replacing them.[69] Many other aspects of health law and policy (such as licensure and reimbursement policies, among other rules) should also play a role in assuring humanistic (rather than behavioristic) mental health apps.[70]

LIBERATING CAREGIVERS, OR AUTOMATING CARE?

Health care can be difficult, dirty, monotonous, and even dangerous work. An incipient army of robotics and AI may lessen each of those burdens. Robots have already entered hospitals and nursing homes. Few question their deployment as cleaners, prompters (to take medication, say, or to avoid salt), lifters (of immobile patients), or in various other helping roles. All free up caregivers—whether relatives or nurses, doctors or friends—to take on other, more meaningful interactions. The Japanese Robear can lift patients in and out of bed, sparing some aides and orderlies back injuries from lifting. Automated pillboxes may prevent medication errors, a bane of geriatric care.

Ideally, such innovations free caregivers to engage in interactions that are uniquely human. Consider, for instance, this profile of a hospice worker and the types of comfort she offers to the dying:

> Heather is not brisk or efficient, as nurses in hospitals are. She is purposely inefficient, in fact. Most of the time when she visits patients, she doesn't have much to do: she takes vital signs, she checks that there are enough supplies and medications in the house, she asks if old symptoms have gone away or new ones developed. . . . But, even when there's nothing else to do, the idea is to be around longer, to chat, to sit close by, to put her hands on the patient's skin as she goes about her checkup. Her visit may be the high point of the day for the patient. . . . So Heather moves slowly; she sits down; she delays; she lingers.[71]

This is a type of care that is all too rarely found in time-pressed health care settings. Even when those sympathetic to nurses develop minimum staffing ratios, they rarely factor in time for this type of purposeful "inefficiency." That is a mistake, because the ineffable aspects of this hospice nurse's job are among the most important.

Some AI and robotics are now designed to take on these tasks of companionship and connection. Reporter Nellie Bowles has described Bill, an elderly man who was overjoyed at the digital ministrations of Sox. Sox was little more than an animated cat on a tablet screen that asked him questions about his day, responded to his queries, and occasionally nagged him to eat better.[72] Bill knew that distant workers were operating Sox, typing in what it should "say" as they watched his movements. But he found that even its cursory questions about his life and health helped relieve the loneliness that had afflicted him since his wife died. As another Sox client stated in a video, "It's nice to have someone ask how your day was, and, you know, if something's bothering you."[73]

An Irish robot companion, Mylo, can also adopt a feline face as its own.[74] Mylo is a mobile object, with a frame reminiscent of C-3PO of *Star Wars* fame. The social entrepreneur who developed Mylo has marketed the robot as both comforter and minder, capable of reminding dementia patients to take their daily pills or to alert family when its client has, say, screamed for help or remained immobile for an unusually long time. Users can rent Mylo for only nine euros a day, which is less than the cost of an hour of hands-on, in-person care. That cost differential threatens to push social and care robots beyond their proper role of helping human caregivers and toward substituting for them.

It is easy to project personality onto dolls, cartoons, and televised characters. For a sufficiently isolated person (think of the Tom Hanks character in *Cast Away*), even a volleyball may suffice. The media theorists Byron Reeves and Clifford Nass have observed that many people tend to relate to media personalities as if they were actual presences in their lives. These "responses do not require strategic thinking; rather, they are unconscious. The responses are not at the discretion of those who give them; rather, they are applied always and quickly."[75] Robots can provoke similar affection, inviting their own anthropomorphization or zoomorphization.[76] When they evoke those feelings of trust or friendship among vulnerable populations, sensitive ethical issues arise.[77]

Consider, for instance, potential applications of animal-like robots to substitute for pets. Animal-assisted therapy has been very popular in some long-term care facilities, like "green nursing homes."[78] Unfortunately, animals can be hard for staff to handle, and many institutions fear liability if animals bite or scratch. Robots promise a safer alternative. Researchers in Japan have developed Paro, a robotic stuffed animal that looks like a baby seal, as a companion for dementia patients. Paro can act as a kind of pet, mimicking responses of appreciation, need, and relaxation. It bats its eyes, mewls and squeaks, and can jiggle its whiskers and tail. It attempts to provide nonhuman companionship without the risks of animal contact.

Massachusetts Institute of Technology researcher Sherry Turkle has urged greater caution as we become increasingly reliant on robotics. Her close observation of vulnerable populations reveals just how profound the impact of simulacra can be: "Elders in a nursing home play with the robot Paro and grapple with how to characterize this creature that presents itself as a baby seal. They move from inquiries such as

'Does it swim?' and 'Does it eat?' to 'Is it alive?" and "Can it love?'"[79] Turkle worries about a society in which busy adults leave their aging parents with an array of sophisticated toys instead of visiting them. As founder and director of the MIT Initiative on Technology and Self, she sees excessive engagement with gadgets as a substitution of the "machinic" for the human—the "cheap date" of robotized interaction standing in for the more unpredictable but ultimately challenging and rewarding negotiation of friendship, love, and collegiality. While videos and studies document Paro's positive effects on the mood of the chronically lonely, Turkle suggests that the innovation may just excuse neglect. Why visit Grandma, some might rationalize, when a robotic animal companion is available?

Paro's promoters point to the practical need for this type of innovation, given the loneliness of many elderly people. Even pet therapists can only visit for a few hours at a time. If there really is no alternative, no human or animal available to show concern or affection, isn't Paro better than nothing? To the extent that the "ages of man" come full circle to infancy in dotage, isn't Paro merely a high-tech version of the *Velveteen Rabbit*, that classic children's tale in which a stuffed animal becomes real for a lonely boy? Paro isn't substituting for real companionship generally, its defenders argue, but only for a small segment of the population whose care needs could easily overwhelm private means and public coffers.

Proselytizers of social robots in elder care tend to cite statistics on the shortage of quality nursing home care, and the high cost of such services, as an economic imperative for their work.[80] In many developed economies, the age pyramid is turning into a rectangle, with as many elderly as there are young people. The imminent retirement of millions of baby boomers in the United States has been a concern of long-term-care policymakers for decades. Aging populations are assumed to create enormous demand for social-care robots. This plays into a substitutive mindset, where technology stands in for humans. Indeed, set within the politics of "build a wall" xenophobia or ecofascist nationalism, technology may provide an excuse for rich countries to restrict immigration. For example, an aging population may decide to shut out younger workers from abroad, assured that robots can do work now completed by immigrants. This move for self-sufficiency may be self-defeating: as long as robots (or their owners) are taxed less than labor, they are less capable of paying for old-age pensions and health care than the workers they replace. Policymakers should explicitly address such long-term fiscal effects before they accelerate the automation of care.

A big picture perspective on the political economy of caregiving is critical to fair evaluation of the scope and potential of social robots. Funding shortages for elder care are not natural dictates of economic science. Nor is the workaholism that can make more frequent family visits seem like an unattainable luxury. They are artifacts of specific public policies, and they can be reversed (or at least ameliorated) by better policies. If wealthy countries had more open immigration policies, workers from many other nations would take on caregiving work. As Ai-Jen Poo has demonstrated, humane arrangements for immigrant labor (and paths for citizenship) are paths to global solidarity and mutual aid.[81] Similarly, firms lamenting a lack of qualified workers have been swamped with applications when they raised compensation levels.[82]

Admittedly, different governments may make different judgments about the proper extent of robotic care for the elderly. There is no universal prescription here. Japanese leaders have been particularly enthusiastic about deploying technology in health care settings.[83] Several commentators have argued that Shinto and Buddhist traditions make Japanese citizens more accepting of robotics and allied technology. Former MIT Media Lab director Joi Ito, for instance, discerned a deep connection between forms of animism and acceptance of robots.[84]

Japanese culture, however, is not monolithic, and its pluralism includes voices expressing caution about the spread of AI.[85] For example, in a vivid exchange, animator Hayao Miyazaki dismissed an AI animation of a grotesquely humanoid figure:

> I am utterly disgusted. If you really want to make creepy stuff, you can go ahead and do it. I would never wish to incorporate this technology into my work at all. . . . I strongly feel that this is an insult to life itself. . . . I feel like we are nearing to the end of times. We humans are losing faith in ourselves.[86]

Notes of resistance like Miyazaki's complicate any simple narrative of unified Japanese acceptance of robotics. For every enthusiastic adoption of, say, caring robots, there are also stubbornly insistent defenses of the human, of personal contact, and of traditional modes of interaction. We must treat "cultural" arguments with caution, because presentations of a unitary cultural viewpoint tend to marginalize or ignore altogether subcurrents of dissent and concern.[87] Just as Amartya Sen challenged Singapore's leaders' promotion of distinctively "Asian values" as a bulwark against human rights claims, we must today resist facile assumptions about cultural homogeneity.[88]

While movies and manga may be held up as evidence of Japanese culture's devotion to automation, there are notes of tragedy and regret, too. The prophetic (if blunt) 1991 Japanese film *Roujin-Z* (Old Man Z) expressed grave anxieties about substituting machines for human caregiving. In that film, a suffering and infirm elder is "cared for" by a hyper-mechanized bed that anticipates his every need—to watch television, have his nails clipped, be bathed, even to eat and to defecate. It is not simply that the film's presentation of totalized, mechanical care is grotesque. The very idea of such a machine, divorced from human agency, guidance, and dignity, is deeply troubling. The stakes are so high, the concept of what makes life worth living so ineffable and variable, the decision-making so intricately shared among care givers and care receivers that routinization seems ill advised at best, and deeply offensive at worst.

To be sure, each of these examples may exemplify "cultural dissent" in Japan, rather than typical attitudes. Danit Gal, project assistant professor at Keio University's Cyber Civilization Research Center, has helpfully typologized attitudes toward AI and robots on a spectrum between "tool" (to be used instrumentally) and "partner" (a designation opening up a broader range of empathy, solidarity, and even friendship or love).[89] Gal offers several indications that Japan is closer to the "partner" side of the spectrum. But far from concluding that this indicates a monolithic set of "Asian values" toward technology, Gal argues that South Korea has established "a clear human-over-machine hierarchy, where AI and robots are expected to support and further enhance this human dominance." She also concludes that China is in the middle of the two paradigms, with ongoing political and cultural debate on the fundamental premises of human-robot interaction. As these distinctive approaches evolve, the economics profession should not presume to put a finger on the scale in favor of substitutive automation. It can just as readily prescribe human-led care.

THE POLITICAL ECONOMY OF ROBO-CARE

Larger trends in workforce participation and demand also matter. Robotic caregiving makes far more sense in a society where the adult children of the elderly are under constant pressure to work more; an overburdened "sandwich generation" has to sacrifice somewhere. If, by contrast, productivity gains were better distributed, demand for robots in elder care would likely diminish. The diffusion of innovations is due less to the existence of the device itself than to the need it serves in a certain *sociotechnical system.* As Martha Fineman has argued, we can redesign social systems to change these incentives and pressures, rather than trying to change people to fit into increasingly dysfunctional systems.[90]

Finally—and most importantly for my argument here—a choice between social robots and health care workers presents a false dichotomy. Even ardent supporters of social robots now tend to portray them as adjuncts to, rather than replacements of, caregivers. Evidence backs them. For example, deploying a Paro in conjunction with intensive engagement by nurses and other human caregivers offers better results than more hands-off approaches. Over the past decade, researchers have conducted a number of studies about such uses of the robot, with many positive outcomes. Psychiatric researchers in Taiwan found that "robot-assisted therapy can be provided as a routine activity program and has the potential to improve social health of older adults in residential care facilities."[91] Norwegian planners found that "Paro seemed to be a mediator for increased social interactions and created engagement" when elderly patients with dementia were exposed to it.[92] Human caregivers can structure interactions with social robots to encourage self-expression, concern for peers, and conversation. This complementarity helps both patient and caregiver.

Both professional schools and health care systems should cultivate the skills of professional mediators between vulnerable individuals and technical systems. As ethicist Aimee van Wynsberghe has argued, a robot's deployment reflects a vision of a practice—here, the practice of caring. In many settings, "human touch, eye contact, and human presence" are essential to caregiving. When a robot can enhance such encounters, it can be a welcome part of values-centered design.[93]

Advanced health care systems are already embracing this complementarity. In the Netherlands, the documentary *Alice Cares* followed the experimental deployment of a companion robot to supplement visits from home health aides. The resulting interactions were, by and large, positive. One elderly woman showed the robot photo albums; another watched football with her. The robot would cheer, nod, and occasionally speak encouraging or concerned expressions. Home health workers assured the elders that the robot was not meant to replace human interactions, but to enliven them. That was a predicate for trust.[94]

Developing trust is necessary, given widespread concerns about the deployment of autonomous machines. According to a Eurobarometer survey, 60 percent of EU nations would ban robotic "care of children, elderly people and people with disabilities."[95] Their views are not mere Luddism. The rise of autonomous robotic "carers" would mark a deep social division: between those who experience human-to-human connection and those relegated to software and machines. It is questionable whether they are even receiving care at all. Care, as opposed to a repertoire of external behavior mimicking care, can only arise out of a mutualistic relationship where the carer is free, at least in principle, to stop caring. It is the continual reaffirmation of a willingness to devote time and effort to another that makes care so precious, and so uniquely the province of human beings with free will. AI and robotics can assist in improving the conditions of care, but cannot do the job themselves.

MAINTAINING THE HUMAN TOUCH IN MEDICINE

We can learn a great deal about the future of robots and AI in medicine from the successes and failures of another critical technology:

pharmaceuticals. No sensible doctor would want to go back in time to practice medicine before the advent of modern drugs. Antibiotics cure pneumonias that would have been deadly less than a century ago. Still, the question of which drug to take, at what dose, and for how long is controversial for many conditions. Robots and AI will be similar. Some doctors will prescribe them widely; others will be more cautious; and we all hope that better data collection will allow objective observers to separate hope from hype.

Once we get beyond the basic questions of the safety and efficacy of automation, even more difficult judgments arise. Were decisions about the nature and pace of AI deployment to be made simply according to a market logic, cheap robots and apps might quickly become the new gatekeepers for access to specialty care—or entirely usurp human doctors for many patients. A personal doctor might be dismissed as a luxury, like a bespoke tailor. On the other hand, were state actors to take too large a role in the pacing of automation, political considerations might freeze inefficiencies in place. Market and state are both best balanced by what sociologist Eliot Freidson called a "third logic"— that of professionalism. In professions, workers with specialized knowledge, who provide especially important services, "have the power to organize and control their own work."[96]

Medicine is one of the oldest professions, but the rationale for its autonomy has changed over time. When medical treatments were a crapshoot at best, the physician was akin to a haruspex, respected for having some intimacy with unknown forces. As science made medicine more reliable, medical boards developed standards to protect patients from quacks and charlatans. There were also critical questions about managing the risks of medical research that were not nearly as pressing in, say, marketing research. To address all those concerns, the profession of medicine is charged with using its privilege (to license or deny qualifications to provide care) to protect the public, and especially the most vulnerable.

Unfortunately, clever marketing for some mental health apps could set the industry on a divergent and unstable path. To potential customers, these apps may be used as a cheap, convenient alternative to therapist visits. That contrasts with the attitude of many regulators (ranging from licensing boards to consumer protection agencies), to whom apps are mere games, diversions, information services, "general health" and wellness aids—anything to avoid traditional duties and responsibilities of therapists.[97] While such regulatory arbitrage may work as a short-term legal strategy for particular firms, it corrodes the trust necessary for the industry as a whole to thrive. Patients also risk acting on substandard advice. For example, when researchers in Australia studied eighty-two mobile apps marketed to those suffering from bipolar disorder, they found that the apps were, in general, "not in line with practice guidelines or established self-management principles."[98]

Doctors Adam Cifu and Vinayak Prasad began a relatively recent book with a shocking finding: "Despite tremendous advances in the clinical, genomic, and surgical sciences, doctors continue to use medical practices, sometimes for decades, that are later shown to be of no benefit to their patients."[99] Prasad has summarized the problem in a statistic: "46% of what we [doctors] do is wrong." To be sure, there are some easy and direct diagnoses, and for most of us, that is our main experience of the medical system. More evidence will help us find more of these "silver bullet" cures, and even to automate them. But we are far from an AI-driven path to guaranteed decreases in morbidity and gains in longevity.

Indeed, AI's and robotics' greatest impact on health may be indirect, thanks to developments far outside the health sector. In recent years, public health researchers have revealed the exceptional importance of social determinants of health—nutrition, sleep patterns, job stress, income, and wealth. Although AI may create more bounty for all, potentially boosting welfare via effects on all of these areas, it can just as readily accelerate unwanted surveillance, social division, and zero-sum competition. These effects can have just as much or more effect on lifespan and well-being as advances in surgical robots or AI-optimized diets.

For most healthy people, doctoring looks like a simple task of pattern recognition (diagnosis) resulting in procedure or prescription. The idealized TV doctor judiciously weighs the evidence, forcefully recommends a course of action, and moves on to the next case. Were things so simple, robots could eventually stand in for doctors. But in the real world of medical practice, this image is long outdated. There is real and enduring uncertainty about the best course of action in many circumstances. And contemporary medicine demands the participation—or at least the understanding—of the patient with respect to a plan of care. These factors complicate and enrich the doctor-patient relationship and make a deeply human element essential. Teachers and students face similar challenges, as the next chapter shows.

3

· Beyond Machine Learners

In 2018, the hashtag #ThankGodIGraduatedAlready started trending on Chinese social media, along with a chilling picture. Students in a high school class had green and red rectangles around their faces, each with an identification number and descriptor, such as "Distracted," "Answering Questions," or "Sleeping." A camera recorded the classroom each second of each day, generating hundreds of thousands of images of every student's face. AI compared the faces to labeled images, parsing each for signs of learning. Some workers had paired a training set with labels such as engaged or distracted, attentive or goofing off. Now, for the thousands of students under watch, there was never a moment off-screen. Reports summarized how much time each student spent each day concentrating.[1]

A massive technology firm, Hanwang, had developed this "class care system" to monitor and motivate pupils. Another firm, Hikvision, marketed AI to record whether students were happy or sad, angry or surprised. It also gamified scoring, generating leaderboard monitors for each classroom, where students could compare themselves against one another. A schoolwide scoreboard encouraged classes to compete to be the best behaved. Any low-ranking class would know exactly which laggards were dragging the group down. Reporter Xue Yujie analyzed this "intelligent education" initiative in 2019 and observed nearly universal condemnation of it by students online. Students were "crushed," cursing the AI eyes.[2]

Some girls at the school who were monitored by Hanwang complained that the facial recognition system failed to recognize them. If they changed their hairstyle or makeup, they were liable to be confused for another girl. The school sent them for multiple face capture sessions. Other students rejected the cameras outright. One said, "I want to smash them." Even in a country accustomed to mass surveillance, a second-by-second record of daily activities provoked outrage. When the Gates Foundation helped fund cognate biometrics in the United States (a galvanic skin response bracelet purporting to continuously measure student engagement with teaching), a public backlash forced the Foundation to backtrack. It is unclear, however, whether widespread critiques in China will have the same effect. A Hanwang executive told Xue that the Chinese government's New Generation of Artificial Intelligence Development Plan helped inspire the Class Care System.[3]

Meanwhile, in at least one university in the United States, a different future for educational technology (edtech) was playing out. After taking a class in artificial intelligence at Georgia Tech, students found out that one of their teaching assistants—known to them only by the online presence "Jill Watson" (JW)—was in fact a bot.[4] JW would post questions midweek and answer students with canned responses. Beg for a chance to revise an assignment? JW would reply, "Unfortunately, there is not a way to edit submitted feedback." JW emailed back promptly, but not too promptly, lest its timing make students suspicious. In response to requests for help, JW would chime in "yes" and "we'd love to" on online discussion boards. Other routine questions about deadlines and homework received helpful (if chirpy) responses.

Students interviewed in a newspaper story on JW seemed pleased by its deployment. They appreciated JW as conscientious and prompt. "I didn't see personality in any of the posts. But it's what you'd expect from a TA," said one student who asked for help on homework. Ashok Goel, the professor of computer science who deployed JW, said the bot could eventually answer 40 percent of the ten thousand or so questions posted by students in the course over a typical semester.

So is this the beginning of the end of teaching assistants in computer science—or even professors themselves? Goel suggests the opposite. With a bot fielding easy questions, TAs are freed up to answer more difficult ones. By focusing the technology on routine questions, Goel and his team are committing to a human-respecting near term, where software primarily aids existing educators.[5]

On the other hand, whatever the intentions of computer scientists, strong political and economic currents will push innovations like JW in another direction—toward replacing teachers and constant monitoring of students. Georgia was one of many states to slash funding for public education in the wake of the Great Recession. The COVID-19 crisis has pressured universities to cut costs further and put more material online. Powerful players in education policy, ranging from globally influential foundations to top level bureaucrats in Washington and Brussels are also fixated on cost cutting. Instead of raising taxes to expand existing universities, California in 2016 deployed a poorly thought out set of online courses to make up for a lack of slots in its colleges.[6] AI (to teach) and robotics (to monitor testing) are a plausible next step, given the need for such courses to monitor students in order to prevent cheating and discipline inattention.

In the United States, advanced surveillance technology already tracks the eyeballs and knuckles of students during online classes and tests. Virginia Commonwealth University has encouraged students to use retinal scans instead of credit cards or cash to pay for meals. As the data trail grows, entrepreneurs aspire to track student movements and more, the better to correlate certain patterns of life with desired outcomes. In other words, Hikvision's incessant surveillance machine is not a strange outlier of "tattletale tech." Rather, it could presage the future of education in an era of crushingly conformist AI.

To put the matter starkly: we have to decide whether we are going to invest in educational AI that is constantly measuring and evaluating students, or if we will focus our energies on products that advance learning in more supportive and creative ways. Limits on a human teacher's ability to answer questions and offer insights are a problem to be overcome; limits on their ability to watch and judge every moment of a student's life are a blessing, a valuable aspect of humane education that needs to be preserved into more technologized futures. Unfortunately, a managerialist mindset has colonized too much of edtech, insisting on the primacy of quantitative measurement. Education has multiple purposes and goals, many of which cannot or should not be reduced to numerical measures. If we let AI turn our focus from actual learning to whatever can be best measured and optimized by computers, we will have missed an enormous opportunity. Worse, we will have allowed technology to usurp and ultimately dictate our values rather than to serve as a tool that helps us achieve them. This chapter examines positive uses of AI and robotics in education, while also emphasizing just how easily they may slip into coarse forms of social control.

THE MANY PURPOSES OF EDUCATION

The path of educational robotics will critically depend on the problems we are trying to solve. Robots and AI are tools, not ends in themselves. As researcher Neil Selwyn has observed, debates over the scope and intensity of automation in classrooms tend to be "proxy battles for much broader struggles over the nature, form, and function of education in the 21st century."[7] Each of the following is a plausible aim for education, but those developing edtech are not advancing all aims equally:

1) Learning verbal and mathematical / logical / quantitative skills and substantive knowledge in history, social studies, the arts, science, and other fields
2) Preparing for a vocation or career, via skills training, or a profession, via the development of substantive knowledge and critical judgment
3) Competing for better opportunities in education and employment
4) Learning social skills and emotional intelligence
5) Preparing for citizenship, including civic engagement and participation in civil society[8]

For technocratic managers, once goals have been specified, the next step is to measure their realization, via scored tests or other outcome measures. Consider the second aim, preparing for work. Social scientists can complete all manner of analyses to determine which schools best prepared students for jobs or college. Employability and salaries can be measured, and there are even crude estimates of on-the-job happiness. Diverse college rankings have emerged, based on a mix of factors. Some name the best party schools. More serious efforts examine the publications and research profiles of faculty. Economistic measures are premised on a tradeoff between students' cost of attendance and their future earnings.[9] Their fundamental logic is unmistakable: students should choose programs that correlate with the greatest boost to potential earnings, discounted against the cost of that education.

This last, instrumentalist view of education has a straightforward economic appeal, particularly if you view labor as a commodity like soy beans or coal, an input ideally cheapened over time. Just as advances in mining technology may make coal cheaper over time, edtech can replace costly teachers and professors to make the workforce cheaper to prepare and thus to hire. Of course, this is not how the measurement school bills its own intent, which is always framed as a way of giving students the skills they need to succeed.[10] But when success is narrowly defined as income and employability, focusing on outcomes opens the door to radically different teaching methods, ranging from virtual cyber-schools to lectures broadcast on YouTube. Tech boosters reason that if we can just agree on the right test questions (for primary and secondary education) and post-graduate success metrics (for higher ed), any new method of teaching is fine, so long as students subjected to it achieve as much as or more than peers taught by persons.

Artificial intelligence methods fit well with the neoliberal emphasis on "learning for the sake of earning," because machine learning is best at optimizing some numerical value (such as income) based on manipulation of thousands of other variables. But this focus on measurement breaks down in the realm of softer, or more contextual, skills, habits, values, and attitudes. Who's to say what test of social skills is best? Where is the multiple-choice test for good citizenship, democratic engagement, or political wisdom? At some point, traditional teaching methods may be *constitutive* of critical goals of education—and, thus, of education itself. To say an activity is constitutive of a practice means that the practice (here, education) does not really exist without the activity (interaction among students and teachers).[11] As excited as we may be about a futuristic hand-held scanning device that could diagnose any physical illness (and even treat some), it is hard to develop such excitement for a teaching robot presented as a replacement for interactions with human teachers. There are vital social and interpersonal skills that children (and even adult learners) need to succeed as workers and citizens.

For instance, workplaces routinely demand interacting with peers and a supervisor. Classroom exercises can embody that type of interaction. In some cases, a student may feel confident and competent; in others, some bravery to try new things will be necessary. To the extent that jobs involve some level of human interaction, it is hard to see all this being simulated via technology.

Nor is engagement an easy problem to solve. I once asked a prominent Google engineer about his experience with online education. He said that the movement for massive open online courses offered excellent information but stumbled when it came to motivation (an observation confirmed by the often high dropout rates in such courses). And given concerns about machine-driven manipulation and privacy invasion, educators probably should not strive to develop machines that exceed human capacities in motivating learners. The case is even stronger for democratic citizenship; collaboration on civic projects is at its core a problem of human relationships.[12]

Edtech enthusiasts who want to replace professors with software are also quick to forget one of the most important roles of the university in contemporary society: to provide disinterested analysis of topics ranging from philosophy to history, art to computer science. There is a real synergy between great research and great teaching that allows thinkers on the cutting edge of their fields to understand the history of their expertise and to communicate that expertise to students. Further breaking apart the traditional unity of teaching and research risks a rupture between the knowledge-creating and the knowledge-disseminating parts of the university. To promote truly engaged scholarship, we need researchers who are obliged to explain the relevance of their work to students and the public at large and at least some critical mass of teaching faculty who are contributing to knowledge rather than focusing solely on making it accessible to students.

THE SIMPLE NEOLIBERAL VISION

The history of education policy has featured an enduring conflict between traditionalist, vocational, and more open-ended and experimentalist visions of teaching. For traditionalists, education's primary purpose is to bind the generations with a common (if evolving) experience of the "best that has been thought and said."[13] Pragmatic modernizers have pushed back against traditionalists, emphasizing the vocational need to train students for jobs. Behaviorists complemented this drive for practicality by offering drilling methods meant to rapidly inculcate knowledge. Experimentalists cautioned against that approach. Given how rapidly social conditions changed, they warned, it would be unwise to put too much stock in particular substantive knowledge—whether learned traditionally or in tech-enhanced computer drills. Instead, learning how to learn is the key skill students require. Experimentalists also recognized the importance of local variations in priorities and emphases.

Unfortunately, top-down "prescriptions for excellence" are now all too common among neoliberal policy elites. Their rationalization of education as "workforce development" fuses vocational and behaviorist tendencies. It could become far more granular and manipulative once accelerated by AI. One way of mapping out an ideal educational system is to project backward from the demands of the employers who will eventually be hiring students. Each year, a certain number of jobs in art, design, programming, management, and dozens of other broad categories are either filled or go wanting. Graduates from various educational programs either find positions or suffer unemployment. From this perspective, education policy is a simple matching problem. If jobs for programmers are not being filled, deans should get more computer science majors into classrooms or set up "coding boot camps" to drill tech skills into English majors after they graduate. In the future, intelligence and psychometric profiles could match individuals to the positions that most closely match their own skills and dispositions. AI recruiters are already screening applicants for cultural fit with firms, and algorithmic systems sort résumés. Influenced by such automated hiring practices, schools could reengineer courses and activities to give students the skills and attitudes to best fit into jobs where they are needed.

To the extent leaders want workers of the future to be bossed around by robots or algorithms, it would make sense to rout them into schools with mechanistic lesson plans and highly controlled teachers—culminating in robotization of the educational workforce. This may sound bleak, but according to some sharp critics of mainstream educational systems, it would be the natural culmination of the worst aspects of schooling. For example, Nikil Goyal has argued that standardization congealed in the early twentieth century, when factory labor and repetitive office tasks became a template for standardized classrooms.[14]

Education expert Audrey Watters has explored how behaviorist paradigms of education have come to inform a "Silicon Valley narrative" of higher education.[15] She has excavated extraordinary documents and plans from the early twentieth century. The first US patents for "teaching machines" were issued over a hundred years ago. Sidney Pressey, a psychologist at Ohio State University, achieved notoriety for developing an "automatic teacher" in 1924.[16] The Pressey Testing Machine let students choose one of five options and offered immediate feedback as to whether the choice was right or wrong (via an indicator at the back that recorded the number of correct answers).[17] Each press of a button (or answer) would advance a mimeographed sheet further along, to reveal the next question. The machine effectively Taylorized the administration of multiple-choice tests and freed the teacher from the labor of having to mark each one individually.

The Harvard psychologist B. F. Skinner pushed the behaviorist model of the Pressey machine even further. Skinner rose to fame by developing the mental models of pioneering psychologists like Pavlov. Pavlov's famous dog learned to associate the ringing of a bell with feeding, and thus salivated (and presumably had some degree of joyful anticipation) whenever his master rang a bell. Skinner believed that a similar pattern of stimulus and reward drove human behavior as well. And by the mid-twentieth century, he was convinced that "the simple fact is that, as a mere reinforcing mechanism, the teacher is out of date."[18] His "teaching machine" let students pull levers to record their answers and turned on a light when a correct answer was registered. Some versions of teaching machines could even dispense a pellet of candy once a student had answered enough questions correctly.[19]

We may laugh at the crudeness of Skinner's mechanical vision of the learning process. But such simple mechanisms have had powerful impacts in other settings. Testing of newsfeed items, search results, and numerous other aspects of online life, for example, is ultimately behavioristic. Few at Facebook or Google care about why you click on an ad; their goal is simply to ensure that you do, based on thousands of experiments and automated observations. They follow in the footsteps of designers of slot machines, who have closely studied how spinning reels and images mesmerize gamblers, leading them to spend more and more "time on machine."[20] Gamifying learning may be less a matter of playfulness than of profit.[21] User experience specialists admit that this quest for "stickiness" is a general principle of internet design.[22] And it is easily applied to robotics, since humans have a more visceral connection to moving, lifelike objects than to stationary computers.[23]

AUTOMATING EDUCATION

Children, with their developing brains, are extraordinarily adaptive. Given the expected pervasiveness of AI in future societies, should children be exposed to it at a very early age? If so, how much? And can it ever substitute for human concern and connection?

Perhaps inspired by these and similar questions, Tokyo University researchers have tested Saya, a remotely controlled humanoid robot, as a teacher of classes in both elementary and university contexts.[24] Saya was designed with an animated face to convey emotional responses to student answers and behavior. The robot had at least seven facial expressions (including sadness, anger, and happiness). Nineteen parts of the face of the robot could move, including "brow raisers," "cheek raisers," "lip corner depressors," and "nose wrinklers."[25] Researchers coordinated the content of Saya's instruction with the expressions its face generated. For example, a correct answer might be greeted with a smile; the order "quiet!," along with an angry face, would confront chatty, distracted students.

The Tokyo University researchers only deployed Saya in a handful of classes, so it is hard to generalize from their results. But they did observe some notable differences between the university and elementary students' reception of the robot. The younger students were more likely to engage, to enjoy the class, and to report increased motivation to learn about science.[26] The older students, however, were more reserved.

The relative caution of the older students suggests at least two interpretations. Perhaps they perceived the instructor behind the Saya as merely phoning it in, using telepresence to avoid the burden of actually being present in the classroom. After more than a decade of interaction with human instructors, the crude expressions and vocalization of the Saya may have seemed too tinny or inauthentic. Saya's face also seemed to fall into that uncanny valley. The robotic appears as one category; the human, another. And when robots come close to human instantiation but fail to achieve it, aversion is almost instinctive.

One wonders, though, if the uncanny valley effect would persist if young people were exposed, from a very early age, to robotic teachers or even caregivers. The question of robotic acceptance in education is not simply one of documenting and accommodating what people prefer in the present. Rather, it requires a future-forming practice, one where children's earliest exposures—to human or robot teachers—could set them on a path toward a lifetime of similar preferences in later grades, and even as workers and consumers. This conditioning effect is one reason why so many firms vie to provide educational technology: the younger a student is, the more malleable and receptive they are to view certain interfaces and advertising as a "natural" way to relate to the world.

The Saya experiment may not seem particularly impressive. The robotic teacher was not autonomous; in some ways, it was little more than a puppet of a human instructor. It only "taught" a few classes.[27] Advances in affective computing could, however, make robotic instructors more versatile, comprehensive, and effective conveyers of knowledge or prods to learning. Education researcher Neil Selwyn has even titled a recent book *Should Robots Replace Teachers?* While Selwyn is skeptical of widespread robotic teachers, he does identify specific instances where something may be better than nothing—that is, where human teachers are not available, some education technology may help promote learning.[28]

Is a revolution in primary education also imaginable, given a critical mass of parents willing to keep kids at home to be instructed by a teaching robot? This radical vision of change runs into some practical difficulties almost as soon as we imagine its implementation. In a high percentage of households with children of grade school age, both parents work.[29] School functions not merely as a place to learn, but as a source of child care—a service that is increasingly expensive. So robotization of primary education won't necessarily save money by encouraging students to stay at home for online exercises. Rather, it could just shift resources from one group (teachers) to another (babysitters). Programming a "nanny module" into education robots may be another option—but that raises its own ethical dilemmas.[30]

Hard-core futurists could promote robotic child care as a complement to robotic teachers. The third-grader of the future might have breakfast served by some modern-day equivalent of Rosie from *The Jetsons*, be ferried to sporting or music events by a self-driving car, and enjoy robotic instruction at school. But all of these developments are a long way off and may well remain so given parents' suspicions of ever more technologized childhood.

Should they, instead, welcome such technological advances? Given the unique vulnerabilities of childhood, the answer is likely no. The insertion of robots into roles traditionally performed by humans can immediately be recognized by adults. When students fail to recognize the robotic nature of an instructor, they are not simply suspending disbelief before an ingeniously humanlike machine. Rather, the robot is essentially deceiving the children into believing that they are being paid attention to by an adult.[31] And the threat here is not simply the deception of children, but rather a subtle indoctrination, a training that the given and the made, the human and the machine, are in a deep sense equal and interchangeable.[32]

The key question is how to balance a healthy enthusiasm, and even affection, for robots, with a sober awareness of their otherness.[33] A robot can observe behavior around it and implement responses to that behavior, but it cannot experience that behavior as a source of meaning or motivation, the way humans experience it. There is also always a residue of the manipulative here, however strenuously technologists try to evade the issue, because the robot is programmed with certain ends in mind. A person relates to other persons as autonomous subjects (with all the privileges to do good and bad, to care and neglect, that such status entails). A machine, even when programmed to have some degrees of freedom, never experiences such autonomy as an agent, if only because our experience of agency is rooted in a wholly distinct embodiment as carbon-based biological creatures rather than silicon-based mechanical ones.[34]

Students should be taught, from as early an age as is practicable, that they are more important than the robotics around them, that these software-based tools are for their good and use. They should respect robots as their or another's property, but should be discouraged from characterizing them as "friends." Better, rather, to consider robotics and AI as a kind of digital environment, to be conserved and domesticated, but without the level of regard and affection we reserve for the fauna and flora of the natural environment. That natural environment is the ground of our being; robotics and AI are only tools.

We should not be distracted from that fundamental reality by Hollywood depictions of robots (so often played by humans) who make some emotional demand for recognition, love, or respect. To be sure, disrespect and abuse of humanoid robots is troubling to the extent it may inculcate a more generalized callousness.[35] Yet such concerns do not dictate anthropomorphization of robotic systems. Rather, they counsel extreme caution in introducing humanoid robots into new settings, particularly among vulnerable populations like children and the elderly.

WHO IS YOUR TEACHING ROBOT WORKING FOR?

What is driving the companies behind the most important education technology? And why should we trust them to make software or robots that reflect our values? Alarm bells are already ringing. The edtech giant Pearson has conducted an experiment on thousands of unwitting college students, exposing some to "growth mindset" psychological messaging in its commercial learning software. It found a mild improvement in motivation, and perhaps that insight will be useful to future software designers. But the colleges had contracted for a learning platform, not human subjects research.[36] What other types of messaging could be seeping into the student-computer encounter? Administrators at all levels of education need to be much savvier about how they contract with edtech firms. They need to negotiate for control of data and for notice about revisions to programs. Without such safeguards, the ethical standards of educators will give way to the rough and tumble "whatever works" ethos of tech firms.[37]

Pearson's experimentation is common in the tech world, which has been less than careful about obeying existing laws protecting children. Criticizing Google's ecosystem, activists have accused both YouTube and the Android operating system of improperly targeting ads at kids and tracking them.[38] As a de facto short-term babysitter for the children of harried parents, YouTube has also fallen down on the job by hosting bizarre and disturbing kids' videos, some of which feature automated remixes of cartoon characters tormenting or abusing each other.[39] YouTube eventually responded to public outcry by promising a youth-focused app curated by humans, not algorithms.[40] That backtracking on automation signals a lesson for educators as well; when it comes to children, "whatever keeps their attention" is not an adequate ethical guide for content creation, however easy it may be to measure that engagement using viewership data.

School systems also need to be careful about using technology as an all-purpose surveillance system. Well-meaning administrators are outfitting schoolkids with Fitbit trackers that count every step and laptops that can track eye movements and tally the seconds devoted to each task and screen. The stakes of this data collection are high. Affective computing aspires to register the import of frowns, furrowed brows, smiles, and other telltale signs of epiphany and befuddlement. Profitable firms have compiled a database of emotions associated with millions of facial expressions. Anyone who has been told to "smile more" to get ahead can intuit the problems raised by a system that is constantly registering and recording the external indicia of emotional states.

One can imagine that given enough investment and time, we might look forward to a future learning environment offering almost limitless AI-generated drills and more creative problems. Robotic teachers would always be at the ready with remedial lessons, too. But there are also untold opportunities for manipulation when that much data is collected from children. Sometimes the interventions are invaluable; Facebook, for instance, can algorithmically detect postings by teens that indicate suicidality and refer them to specialized help.[41] But it's also pitched similar capabilities to advertisers to help them identify kids who feel "insecure" or "worthless."[42] There needs to be a line between efforts to help and teach, and the types of creepy data gathering and stigmatic characterizations that erode trust. Students and parents should always be entitled to opt out of affective computing.

At present, we don't have nearly the insight we need into the collection, analysis, and use of data by firms developing educational software and robotics. The capacity of modern machine-learning programs to record everything a child says and to increasingly personalize responses—to provoke laughter or concern, affection or a sense of guilt—is a new development in student monitoring that should be rolled out slowly, if at all. Before such educational robots are adopted *en masse,* basic questions about their power and purpose must be answered.

THE DOWNSIDE OF CONSTANT ATTENTION

Surveillance tends to offer immediate, concrete gains mixed with longer-term, more diffuse dangers.[43] The promise of robotics in school—as with so much other educational technology—is deeply personalized lessons. Big data analytics might also vastly reduce costs. A firm might provide educational robotics to schools for free, with the long-term aim of gathering data on students and helping marketers and others understand them. The terms of monetization are critical. Is data used strictly in the educational setting, to aid in learning? Or could it inform digital dossiers, secretly stigmatizing students?

Sociologist Erving Goffman developed the notion of "off-stage" spaces, where one can explore alternative viewpoints and behaviors without being watched by others.[44] Poorly regulated robotics systems could ensure that every moment at school is "on-stage." Are educational robots defaulted to record every moment when a child is within earshot? Do they notify children or those around them when recording is happening?[45] Several consumer groups have complained about toy robots spying on the children who used them. According to a petition they filed with the US Federal Trade Commission, "By purpose and design, these toys record and collect the private conversations of young children without any limitations on collection, use, or disclosure of this personal information."[46] Like many terms of service, those governing the robot's interactions with children provided little meaningful protection. They even reserved the company's "right to amend these Terms for any reason if necessary." In other words, whatever protections exist today may well be gone tomorrow, undone by one of the "four horsemen of irresponsibility" I described in Chapter 2.[47]

Biometric voiceprints have become a big business, particularly as a method for authenticating identity.[48] Consumer advocates worry that the voice data is not being properly secured and could be hacked by those with malicious intent. An even bigger danger could come from the processing and use of the data, given increasing pressure on schools and governments to identify "problem kids" at a younger and younger age. According to a recent study, about 20 percent of citizens account for 81 percent of criminal convictions, 78 percent of prescriptions, and 66 percent of welfare benefits—and this group "can be predicted when as young as three years old from an assessment of their brain health."[49] What if certain tones of voice or patterns of interaction are hallmarks of behavioral problems to come? Who has access to such data? And how can they act on it?[50]

There is a strong case for limiting any data collected in an education setting to that setting (barring cases of immediate danger, such as suspected abuse). That is the spirit of California's "Online Eraser" law, which enables children to delete data about their social media presence before the age of eighteen. The European General Data Protection Regulation expands such "rights of erasure" to many other contexts. In time, it should become a global gold standard for the rights of students subject to edtech. Without such protections, why should anyone trust a data-gathering apparatus that has already proven perilous in so many contexts?[51]

Individual families can, of course, try to avoid such tracking and evaluation if they have enough resources, searching out schools with a different approach. Indeed, many parents in Silicon Valley send their own children to low-tech, high teacher-student ratio schools, while building the technical infrastructure for high-tech schooling elsewhere. Yet it is easy to overestimate the power of the individual to resist or exit. Once a critical mass of students is constantly tracked and monitored, those who opt out start to look suspect. What have they got to hide?[52] Grades have become a near-universal disciplinary mechanism; soon, a "total behavior record," measuring students along axes of friendliness, attentiveness, and more, could easily be developed.

MENDING OR ENDING EDTECH SURVEILLANCE

Resistance to such surveillance tends to alternate between reform and revolution, modification and rejection tout court. Some want to improve machine learning in education; others wanted to scrap it. Though the two sides are now united on immediate policy questions, there is a deep tension between the two. Mending surveillant edtech means giving it more data, more attention, more human devotion to tweaking its algorithmic core. Ending it would reorient education reform toward more human-centered models. To decide between the two paths requires deeper examination of both these projects.

No one wants a computer randomly assigning merits and demerits to students and teachers. Some of the girls profiled by the Chinese ClassCare system that was mentioned at the start of this chapter complained that they had been mistaken for one another, especially after they changed hair styles. The technical fix is straightforward, if invasive: including more and more pictures of the students in a reference dataset so their characteristic movements or non-facial features can be spotted.

Even when all students are identified correctly, scholars have raised hard questions about the correlation of facial expressions to internal states of mind. For example, a team of psychology researchers has demonstrated that expressions do not necessarily map neatly to particular emotions, let alone to the broader mental states evoked in evaluations like "paying attention" or "distracted." Since "instances of the same emotion category are neither reliably expressed through nor perceived from a common set of facial movements," the communicative capacities of the face are limited. Moreover, even when we have more data, we must always bear in mind the varying effects of context and culture.[53] Are wandering eyes a sign of day-dreaming or deep thought about the problem at hand? Humans may struggle to make such a judgment, and their own frailties are embedded in the data that feeds machine judgments on the matter.[54]

Undaunted, researchers in affective computing may take two directions at this point, plowing more resources into refining the discernment of emotions or abandoning it altogether. The first approach would double down on the quest for data, looping in unorthodox data sources, such as constant self-monitoring and self-reporting of emotional states by data subjects. Amazon's mTurk researchers have completed more demanding and invasive research costing pennies on the "HIT" (human intelligence task). But is the human heart really so transparent? Why assume emotional self-reports to be accurate, let alone capable of extrapolation to those with similar expressions or deportment? If research subjects are unpaid, they may simply rush through the exercise. If they are paid, self-reports may be little better than rote CAPTCHA filling, where hard-pressed pieceworkers are more concerned about reporting what most other persons report, rather than some "ground truth" of what they see.

The second "answer" to complaints about the accuracy of emotion and attention recognition is to bypass them altogether. A school may simply aim to cultivate the expressions and deportment best correlated with what the school is ultimately trying to produce. The crudest version of this approach might be to map forward detected expression and deportment to test scores, developing some database of outward behavior that correlated to exceptional performance. Administrators could notify parents about how closely their children managed to match behavior that had been determined, in the past, to correlate with high scores. In higher education, starting salaries or degree premiums (how much more graduates earn than others at their age level) may be the target.

There are multiple problems with such a strongly behaviorist approach, unmoored from any concern about what is going on in students' mind. Affective "datafication" of education is creepy enough when it presumes to prescribe moods and attitudes.[55] But at least it gives students some sense of what kind of affective life authorities believe they should pursue or perform. A purely behavioristic model gives up even on that scrap of intelligibility, leaving its subjects adrift as mere mimics.

At this point, one has to wonder whether the game is worth the candle. Everyone needs some time "off stage," unmonitored (or at least far less lightly monitored than the students under the watchful camera of ClassCare or teaching robots). Attention and emotion-evaluation software encourages excessive self-instrumentalization. We all must balance the ways in which we either express or moderate, cultivate or suppress, our emotional responses. This education of the sentiments can be grueling in youth. There is a delicate balance between what Hannah Arendt calls the "protection of the child" and the "protection of the world."[56]

To ensure that children better fit in with the world they will eventually enter, schools shape their students, but ideally only to a point. When control becomes too complete, and particularly when it is imposed in an unreasoning and mechanical way, stresses on the psyche may be enormous. When every act of resistance or relaxation of purpose may be penalized, spontaneity ebbs. Alternately, students may try to instrumentalize their reactions so completely that there is no space between their own desires and responses and those expected from increasingly computerized authorities. However great the merit of social harmony and cohesion, such frozen order is an affront to freedom and judgment.

POSITIVE OPTIONS FOR ROBOTIC HELPERS

If regulators can tame surveillance capitalism's omnivorous appetite for data and control, robotic assistance could play a positive role in many classrooms. Shifting the frame from students controlled by technology to students themselves controlling and playing with it is critical.

Rather than simply herding students from one digital environment to another, inspiring educators show them how to influence and even create such environments. Seymour Papert, a distinguished education researcher at MIT, offered an early version of this type of participation in the 1970s. He developed programs that helped students teach themselves to learn how to program. For those who are spatially inclined, Papert's robotics may be a godsend, a way of realizing, very early in one's life, the power of computer languages to communicate with a mechanical world and to conjure it to do things.[57]

All too often, poorer children are tracked, early on, into a behavioristic model of education, and they experience the school as a site of bureaucratic control.[58] The more fortunate have a more varied cognitive diet and ample opportunities to explore the skills they enjoy cultivating, be they verbal, quantitative, social, athletic, or otherwise. These opportunities should be democratized, and technology can help. Consider the Dragonbot developed at MIT. It is a deceptively simple mashup of a stuffed animal (a cute dragon) and a smartphone (which serves as the dragon's face). The robot can sense students' interaction with a tablet device—it has software to monitor what words or objects on the tablet the student chooses to touch. An internet connection can easily permit instantaneous feedback from the tablet to the Dragonbot, and vice versa.

The Dragonbot can project lessons to children and assess their answers and reactions. It can also turn the tables, asking to be taught. For example, in one scenario, the robot asks students which of five words on their tablet's screen corresponds to the word "dragon," and it can offer appropriate encouragement when the answer is right.[59]

Some might believe that either a skilled teacher or a loving parent should be engaged in all such learning exercises with the young student. And one hopes both are present for much of the student's life. But children also need breaks from adults, to develop their own sense of autonomy and control. At its best, the Dragonbot can give those opportunities. Smart, fun toys can dissolve the boundary between work and play. They can also spark positive interactions, becoming a common focal point for the attention of adults and children.[60]

Learning through teaching is a recurring theme in educational robots. For example, researchers at the Computer-Human Interaction in Learning and Instruction Laboratory at a French university told children to teach a robot how to write letters properly.[61] Programmers intentionally designed the robot to generate poor imitations of letters at the outset so that the children could take pride in teaching it to do better. A robot could inspire students and reinforce their prior learning. Well-designed educational toys can give children a sense of empowerment over (and through) robotics.[62]

Robots may also promote a sense of kinship, friendship, and affection for machines—and this is where the picture becomes a bit more complicated. What is the proper attitude for a child to adopt toward a robotic teaching assistant? Images of children embracing or looking expectantly at their robotic "pals" are both familiar and jarring. Take the smartphone out of the Dragonbot, and it's just another stuffed animal doted on by a child. Older toys, however, were not capable of parsing children's facial expressions second by second and adjusting output in response to that sensory input. The robotic teaching toy occupies a liminal space between subject and object, and we have to decide—as parents, educators, and policymakers—how to define and treat it.

This definitional work is hard to do well; this is one reason why fanciful creatures like dragons (or Pokémon) are better "vectors" for robotic lessons than humanoid robots. One very practical and simple lesson might be to teach children not to be abusive. Korean researchers designed Shelly, a robotic tortoise, to blink and wave its arms when children petted it. But if a child hits it, Shelly suddenly withdraws into its shell and goes silent. Part of a growing field of biomimetic robots, which act like creatures found in nature, Shelly suggests one ready-made model for robot-child interactions: the pet. Mechanical turtles, cats, and dogs (remember the Aibo?) could substitute for children's companion animals, just as the Paro robotic seal did for frail elders. MIT robot ethicist Kate Darling has proposed that "treating robots more like animals could help discourage human behavior that would be harmful in other contexts."[63] For example, a student might learn the basics of an ethics of care around a robot that needed some interaction (or even a charging station) to function. The Japanese game device Tamagotchi capitalized on this sensibility in the 1990s, egging users on to press buttons in order to "feed" or otherwise take care of a digital pet. Starting from that low bar, robots have better emulated animals over time.

Cautious parents might also teach their children to treat unknown robots as feral animals, to be left alone.[64] The distinction between the living and the mechanical should also be made clear. Even Shelly the Turtle risks misleading kids about how to treat a real reptile, setting unrealistic expectations about the tractability of the environment to human desires.[65] In her book *Alone Together*, Sherry Turkle was taken aback at some children's jaded dismissal of a set of turtles sitting nearly motionless on display at a museum. "For what the turtles do, you didn't have to have the live ones," said one girl. Some said they would prefer an active, robotic version of the animal. "If you put in a robot instead of the live turtle," Turkle then asked, "do you think people should be told that the turtle is not alive?" The children demurred. Turkle worried that they preferred an entertaining illusion over dull reality.[66]

The animal analogy also falters when animal-shaped robots mimic human traits. Legal scholar Margot Kaminski worries that robots may engage in "deceptive anthropomorphism" to fool their interlocutors into thinking machines have feelings.[67] Clever marketers can cloak a marketing pitch in the deceptive raiment of a heartfelt appeal from an artificial "person." The kinds of robots we are likely to see in mass use over the next few decades will not be the products of millions of years of evolution. By and large, they will be developed by corporations with a particular business model. Sometimes that business model will follow that of dominant internet firms, where we pay for many forms of content by generating a data trail valuable to marketers, future employers, insurers, and others. A child's dog will not be reporting on the child's behavior in an application for college admission. A robot might.

Young children would find it difficult, if not impossible, to keep such risks in mind when interacting with a robot. And if edtech companies insist on investing in humanoid robotics, despite all the warnings given above, they need to deeply consider their "hidden curriculum"—that is, all the subtle and implicit lessons and biases they are effectively transmitting. Making the default voice of assistants like Siri and Alexa a woman's could be reinforcing outdated and discriminatory gender roles. Why assume helper robots are women? Representations of race in technology also need input from the communities represented—and, ideally, should be led by them. "Nothing about us without us" is as good a motto in robotics as in media, and raises deep concerns about the lack of diversity in many technology firms. Firms must avoid the Scylla of negative stereotypes and the Charybdis of "model minority" idealization. There is little evidence that leading AI and robotics firms are anywhere near capable of sensitively addressing these issues.[68]

COMPARED TO WHAT? EDUCATION TECHNOLOGY AND LESS DEVELOPED COUNTRIES

It should now be clear that schools dominated by AI and robotics would fail to promote several core aims of education. But what if no human teachers are available? So far, I have examined robotics and other educational technology deployed in contexts that are industrialized and at least middle income, by global standards. My main comparator has been the traditional American, Chinese, Japanese, or European classroom, staffed by a well-educated and competent teacher. That type of education is not universally available, whether because of real resource constraints or incompetent governance. While pockets of excellence are available to the rich, poor children in less developed countries are often priced out of quality education. In some areas, there is zero public funding for teachers.

This inequality inspired Nicholas Negroponte's campaign for "one laptop per child." And they could easily spur philanthropic endeavors to push educational robots to classrooms (or poor households) around the world. Neuroplasticity at an early age is an extraordinary opportunity—for example, learning a language is far more difficult after childhood. Even crude robotics, with some basic packages of language and mathematical skills, could be an extraordinary gift to children around the world.

Yet we should acknowledge the limitations of such interventions, too. Though technology is billed as a way of leveling the playing field, it often tends to exacerbate existing disparities. Drawing on research on the diffusion of innovations, computer scientist and international development researcher Kentaro Toyama emphasizes that there are always more and less cutting-edge versions of education technology. In a provocative passage, he notes:

> In any world politically like ours, wealthy, influential parents will secure the best hardware for their own children, while the children of poor, marginalized households will have access to older models in need of repair.... Yet again, the technology will amplify the intentions (both explicit and implicit) of the larger society. And the same will be true of gamified e-textbooks, humanoid teaching robots, or any other novel technology.... If you're interested in contributing to a fair, universal educational system, novel technology isn't what will do the trick.[69]

Toyama's examples, from virtual reality to humanoid robot teachers, may seem fanciful. But telerobotic teachers are being tested, and could easily be marketed as a "solution" to global educational disparities.[70]

Toyama's insights suggest caution about deploying educational technology to classrooms in less developed countries. There is already widespread concern that donations of clothing to Africa from the developed world undermine local industry.[71] How might the robotization of teaching for low-income students reinforce cultural norms that devalue instruction as a profession? How might the lessons chosen for children in India, by firms or nongovernmental organizations (NGOs) in Palo Alto, London, or Washington, DC, diverge from the concerns of the local community? Whose vision of history or politics will best inform civics classes?

Until we have established resilient governance structures to develop just and fair answers to all these questions, we should be cautious about transnational investments in the automation of education. Indian officials have already rejected Facebook Basics, which offered free internet access in exchange for Facebook's control over which sites could be accessed. They believed that that particular "something" was worse than the "nothing" millions of poor citizens had in terms of online access. The same dynamic may play out in education technology.

RESISTING THE NEW BEHAVIORISM

Frederick Taylor revolutionized manufacturing by applying time-motion tests to assembly lines. Each worker's movement was recorded, measured for effectiveness, and either commended or marked for correction. The Taylorist dream of a maximally effective workforce proved a prelude to robotization. Once a single perfect way of doing some task was found, there was little rationale for a human to continue doing it. Taylorism dovetailed with the psychological school of behaviorism, which sought to develop for human beings a blend of punishments and reinforcements reminiscent of animal training.

The rise of data-driven predictive analytics has given behaviorism new purchase. The chief data scientist of a Silicon Valley e-learning firm once stated, "The goal of everything we do is to change people's actual behavior at scale. When people use our app, we can capture their behaviors, identify good and bad behaviors, and develop ways to reward the good and punish the bad. We can test how actionable our cues are for them and how profitable for us."[72] A crowded field of edtech innovators promises to drastically reduce the cost of primary, secondary, and tertiary education with roughly similar methods: broadcast courses, intricate labyrinths of computerized assessment tools, and 360-degree surveillance tools to guarantee that students are not cheating.

In its purest form, the new behaviorism has already generated some notable failures. Some "online charter schools" for children have had laughably bad results, with some studies showing *no* learning in math occurring over a 180-day period.[73] A once-celebrated partnership between Udacity and San Jose State University in California also ended badly, with much higher failure rates than ordinary courses. Complaints about privacy violations and excess regimentation are rife. "Virtual charter schools" have taught at least three hundred thousand students in the United States, promising online content for home schoolers.

But the track record of many is extraordinarily poor.[74] In some instances, 180 days of "learning" at cyber-charters was equivalent to *zero* days in a regular classroom—in other words, there was no educational attainment whatsoever. Similarly, many online postsecondary institutions are dogged by poor job placement rates, spotty instruction, and, at worst, lawsuits accusing them of simply being scams.

As Audrey Watters explains, these failures should not be a surprise. Even in the 1930s, edtech evangelists were proclaiming that radio classes would replace teachers en masse, finally enabling a science of education. Behaviorists predicted that learning machines could radically enhance the efficiency of teaching (recall the candy dispenser mentioned above). For years, Watters has compiled an encyclopedic array of edtech dreams (and their predictable dashing), ranging from virtual reality classrooms to "Matrix-Like Instant Learning Through Brain Stimulation."[75] Though there is little evidence of actual learning advances from these interventions, the promise of some killer app to disrupt education is irresistible to investors, who have poured billions of dollars into the field.[76]

At some point, some of these interventions will probably succeed for some students. But who will be a reliable guide to them? Some edtech industry-funded publications produce their own guides, and the industry aspires to its own forms of certification and online review systems. There are some deep problems with this model. First, deciding what works is a complex process, all too prone to being manipulated.[77] Second, the same investors may be backing both the edtech and those evaluating it.[78] That situation sets up an inherent conflict of interest, biasing the research.

Similar commercial pressures may be leading to a premature plunge into online education. Based on the success of Khan Academy, enthusiasts breathlessly project that a college education can be had for a fraction of the current tuition costs; just record all the lectures, automate evaluation of students, and invite the world to attend, with grades recorded on a blockchain transcript. Get stuck on a lesson? Just keep interfacing with a keyboard, camera, and perhaps haptic sensors. Or perhaps instant message some reserve army of tutors via digital labor platforms like Mechanical Turk or TaskRabbit.[79] Want to prove that you aren't faking exams? Just let cameras record your every move and keystroke—perhaps your eye movements and facial expressions, too.

Beyond the creepiness of such surveillance, there are some obvious incoherencies in this vision. It is far from clear how to replicate the experience of seminars, extracurricular activities, or internships online. Cost-cutters' exasperation at traditional residential colleges is driven far more by an ideology of austerity than a commitment to improving learning. Just as problematic in the automators' visions of viducation is how deflated its take on the production and dissemination of knowledge is. There is no single, final version of a philosophy, history, or psychology course; these subjects evolve, sometimes rapidly, in response to research, scholarly dialogues, and changing times. I may not like the rise of experimental philosophy or rational-choice modeling, but I respect conditions of plurality that let autonomous departments of philosophy and political science plunge forward in those directions. The alternative is a parody of paideia: some central administrative authority parceling out lecture minutes to metaphysicians, utilitarians, and deontologists, likely based on big-data analysis of starting salaries for past graduates exposed to each.

Nor should the role of human insight in science, technology, engineering, and medicine courses be discounted. Even a course as seemingly algorithmic as the introduction to computer science can shift fundamentally thanks to innovative instructors. At Harvey Mudd College, for instance, professors made deep changes in their approach in order to address the field's chronic gender gap. This is a type of progress that arises out of personal interactions among scholars and between teachers and students.

HUMANE EDUCATION

The datafication of teaching and learning has contributed to many troubling trends. And yet it would be foolish to dismiss AI in education wholesale. There are extraordinary online instructional materials in nearly every field. Big data can advance the science of learning as analysts study the most effective ways of conveying material. Students who could never dream of accessing college courses can now watch or listen to their lectures and access self-correcting exercises. The real question is how to realize the promise of AI in education.

The answer lies in a reconsideration of the nature of teaching. Like other professions, it is not simply a matter of conveying information. The duties of professionals do not end with an assessment of the relative likelihood of an intervention working, where "working" is defined by a clear, quantifiable metric. Students require some guidance as to how to conduct themselves—as persons, not simply as absorbers and generators of messages. To think otherwise is to put society on a slippery Skinnerian slope: education conceived as stimulus by the powerful and prescribed responses by the rest.

Of course, there will be exceptional cases. Some rural, bullied, or disabled students will want home schooling.[80] The burdens of attending or merely getting to a school may be too much for them or their families. Even online instruction by a teacher may be too expensive. But we still must ask: However much discretion granted by the state to parents to educate their own children (or to postsecondary students to find their own way), are they really capable of exercising it wisely in a Wild West of automated content providers? Parents may be in the same position relative to edtech, as patients are with respect to pharmaceutical claims. Patients do not navigate that terrain alone—they rely on health professionals. So, too, should we expect there to be a lasting and critical role for education professionals in schools and universities.

In his insightful book *Program or Be Programmed,* Douglas Rushkoff laments the distance between the educational hopes for the internet and its grubby present-day reality. For Rushkoff, the increasing domination of digital interfaces creates a stark choice for students. "In the emerging, highly programmed landscape ahead, you will either create the software or you will be the software. It's really that simple: Program, or be programmed. Choose the former, and you gain access to the control panel of civilization. Choose the latter, and it could be the last real choice you get to make."[81] He is particularly dispirited by the cooptation of inquiry by a fast-food mindset. The reflex to "just google it" to find information widens a cultural chasm between those who aspire to know exactly how knowledge is established and organized and those who merely want a quick answer to a question.

For example, Rushkoff observes that "educators who looked forward to accessing the world's bounty of information for their lessons are faced with students who believe that finding an answer on Wikipedia is the satisfactory fulfillment of an inquiry."[82] This is not to say that Wikipedia has no place in education; its entries often feature helpful primary sources. It also offers important lessons in the bureaucratic politics of knowledge production—how volunteers contribute to a massive project and how conflicts over principles are resolved. Learning about Wikipedia is a good way to gain media literacy; reading it uncritically is unwise. And this caution applies all the more to purely algorithmic arrangements of information, such as search-engine results and newsfeeds: they always reflect some agenda, however objective they may appear at first glance.

Rushkoff's complaint sets one agenda for regulating AI and robotics in education: how to avoid problems already evidenced in digital technology. There is widespread worry that students fail to fully understand the technology they are using—both in a nitty-gritty, technical sense, and also from a critical perspective (assessing the goals and financial incentives of technology providers). Despite all the popular talk of "digital natives," evidence of digital illiteracy is rife.[83] Finding new balances between human-led and software-driven classroom exercises (as well as lectures, tests, capstone projects, essays, and exams) demands experimentation and pilot programs.

For example, Lego marketed Mindstorms bricks as early as the 1990s, giving children a sense of the malleability of the technical environment. They may well help students understand robots less as artificial parents, teachers, caregivers, or peers, and more as one useful part of a larger technical landscape. Productive deployment of new learning models depends on humane and distributed automation that balances the insights of diverse domain experts (including current educators) and technologists.

To alleviate economic pressures toward premature digitization, governments should help colleges pay adjuncts and education technologists fairly, while reducing student debt burdens.[84] A similar dynamic should inform policy in elementary and secondary schools, promoting training for teachers so that they can be full partners in the use of technology in the classroom, rather than having technology imposed on them from above. Collective bargaining agreements and legal protections give unionized workers a chance to help shape the conditions of their labor, including the ways in which technology is adopted. Unions have helped teachers gain basic rights on the job and should give them a right to structured participation in the way future AI and robotics are deployed in the classroom.[85]

This ideal of worker self-governance is taken even further in the case of professions. Professionals have long asserted the right and duty to set key terms of their work. For example, physicians have successfully lobbied politicians to recognize their special expertise and fiduciary responsibility and to exercise that authority in order to structure the medical profession. They run licensing boards, authorized by the state in order to determine who is qualified to practice medicine. They also play critical roles in the regulatory agencies that determine which types of drugs and devices are permissible and which require further testing.

Is it possible to imagine Education Technology Agencies (ETAs) in which teachers would play a similarly decisive role in shaping the adoption of educational robotics?[86] Yes. Just as drug regulatory agencies around the world have helped doctors and patients negotiate the complexities introduced by the contested effects of pharmaceuticals, nations could empower ETAs to evaluate and license apps, chatbots, and more advanced edtech systems and innovations. Ideally, it would complement an education profession empowered to help schools and students choose among multiple technological pathways to aiding learning. To guarantee a truly humane future, we will need a robust teaching profession—at primary, secondary, and postsecondary levels—that can keep the tendency toward mechanization, standardization, and narrowing of education in check. Critical thinking will only become more important as an automated public sphere, discussed in the next chapter, continues to spread unreliable, dangerous, and manipulative content.

4

■ The Alien Intelligence of
Automated Media

The basic contours of mass media–driven politics and culture evolved slowly through the second half of the twentieth century. Since the mid-1990s, change has accelerated.[1] Software engineers have taken on roles once filled by newspaper editors and local newscast producers, but at a distance. They choose content and advertisements for constantly shifting, algorithmically curated audiences. The details of such algorithms are secret, but their broad outlines are clear. They optimize some combination of ad revenue and "engagement"—that is, the quantity and intensity of time spent on the site by users. As megafirms divert ad revenues from the legacy media, they have been richly rewarded, becoming some of the largest corporations in the world.[2] Most advertisers want audiences, not necessarily any particular content. Big tech can deliver.

The largest tech platforms are more than economic entities. They affect politics and culture—areas where metrics, no matter how complex, can only fitfully and partially capture reality. Communications scholars have documented many forms of bias on digital platforms, ranging from the gaming of search results to mass manipulation.[3] These concerns rose in prominence during the 2016 US presidential election, as well as the United Kingdom's Brexit referendum—both of which featured deeply disturbing stories about misleading and inflammatory "dark ads," which can only be seen by targeted demographics. Powerful right-wing echo chambers amplified unreliable sources. Politically motivated, profit-seeking, and simply reckless purveyors of untruths all prospered. What Philip N. Howard aptly calls "lie machines" churned out stories with no basis, tarring Hillary Clinton with an endless series of falsehoods, all to score quick profits.[4] Platform managers cared *that* people clicked on the stories, not *why* they clicked on them or whether the information purveyed was truthful. The full extent of manipulation is only now being unearthed and may never be fully uncovered; trade secrecy protects many critical records from public scrutiny, if they still exist at all.

Nor need there be any conspiracy for propaganda, lies, and sensationalism to flourish. For profit-minded content generators, the only truth of Facebook is clicks and ad payments. Political scientists have estimated that tens of thousands of the tweets "written" during the second US presidential debate were spewed by bots.[5] These bots serve multiple functions; they can promote fake news, and when enough of them retweet one another, they can occupy top slots in hashtag searches. They can also flood hashtags with frivolous posts, making it seem as if genuine public concern about an issue like #BlackLivesMatter is merely noise generated by opportunistic marketers.

These crises of the automated public sphere are so numerous and chaotic that they cry out for systemic categorization and for concerted responses. This chapter explores three sets of problems. First, AI that is blind to the substance of communication is an easy target for extremists, frauds, and criminals. Only human reviewers with adequate expertise and power can detect, defuse, and deflect most of these harms. Second, platforms boosted by this AI are siphoning revenue and attention away from the very media sources that are exposing their failures, leaving platforms ever less challenged in their conquest of communications channels. The economics of online media need to shift dramatically to halt this destructive feedback loop. Third, black boxed AI means we may not be able to understand how the new media environment is being constructed—or even whether accounts interacting with us are actual humans or bots. Here, attribution is key, including disclosure of whether an online account is operated by a human, AI, or some mix of the two.

The new laws of robotics can help us understand and respond to all these threats. They also raise the possibility that the automated public sphere cannot be reformed. When so much of the circulation of media has been automated, new methods of structuring communication may be necessary to maintain community, democracy, and social consensus about basic facts and values. As Mark Andrejevic's compelling work demonstrates, automated media "poses profound challenges to the civic disposition required for self-government."[6] These concerns go well beyond the classic "filter bubble" problem, to the habits of mind required to make any improvement in new media actually meaningful in the public sphere.

The new laws of robotics also address the larger political economy of media. The first new law of robotics commends policies that maintain journalists' professional status (seeing AI as a tool to help them rather than replace them).[7] Content moderators need better training, pay, and working conditions if they are to reliably direct AI in new social media contexts. One way to fund the older profession of journalism and the newer ones created in the wake of media automation is to reduce the pressure on publications to compete in a digitized arms race for attention online (a concern of the third law, proscribing wasteful arms races). Later in this chapter, we will explore ways of shifting funds away from mere platforms (like YouTube and Facebook) and toward the persons actually responsible for helping us understand and participate in politics, policy, and much more.[8]

This will be an exceedingly difficult project, in part thanks to new forms of authoritarianism promoted by the automated public sphere. Some state media and internet regulators, even in ostensible democracies, are broken beyond repair. But this just means that the project of protecting media is all the more urgent in jurisdictions that have not succumbed to the authoritarian trend. And even in places that have, citizens who are painstakingly constructing alternative media ecosystems should find much useful in the four new laws of robotics and their elaboration ahead.

The political, economic, and cultural failings of massive online intermediaries deserve notice and rebuke. While such firms often say that they are too complex to regulate, Europe has already shown how one small corner of the automated public sphere—name search results—can be governed by law in a humane and inclusive way. We can learn from that experience to develop a broader framework of regulation for the most pervasive form of AI now influencing our lives—the automation of media. And that framework of responsibility—built on the concepts of *auditing* the data fed to algorithmic systems and *attributing* the actions of such systems to their controllers—should inform the regulation of AI as it enters more of the physical world through robotics.

WHEN WORDS CAN KILL

Automated media may seem like an unlikely or unfortunate topic for regulatory attention. It is only words and images, far from the life-or-death maneuvers of a surgical robot or autonomous weapon system. The bloody consequences of out-of-control dissemination of hate speech, however, are now well-documented. For instance, in India, Sri Lanka, and Myanmar, bigots and bystanders have trumpeted baseless lies about Muslim minorities on social networks. A bizarre rumor that Muslims deployed "23,000 sterilization pills" virally spread on Facebook, leading one mob to burn down a restaurant they thought poisoned food with the pills.[9] The restaurant owner had to stay in hiding for months. Nor are such incidents rare. Weaponized narratives of racial resentment, as well as outright lies, have influenced crucial British, Kenyan, and American elections.[10] They race through large internet platforms, stochastically radicalizing users.[11]

Dangers abound when a critical social role, historically premised on judgment and human connection, is suddenly automated. Managers and engineers have repeatedly optimized algorithms badly, maximizing ad revenue via a steady diet of clickbait, sensationalism, and manipulation. One might hope that view counts on videos are one last stand for human quality control. However, "viewing farms" run by bots have been rampant.[12] Even when views are authentic, mere "engagement," measured by a user's time and typing on a site, is a very crude proxy for time well spent online, let alone effects on democracy or political community.[13]

To be sure, social networks have also connected communities and advanced laudable social movements. But we would not accept pervasive automation in health care or education if it resulted in millions of students failing to learn basic facts about math and science or if it caused hundreds of preventable injuries. It is time to take our automated public sphere more seriously, before sinister actors manipulate it further. That will require recommitment to two simple principles. First, free expression protections apply to people first and to AI systems secondarily, if at all. Massive technology firms should no longer be able to hide behind "free speech" as an all-purpose talisman of irresponsibility. Second, leading firms need to take responsibility for what they prioritize or publish. They can no longer blame "the algorithm" for the spread of harmful information and incitement. We can either demand humane values in artificial intelligence or suffer the consequences of its cultivating inhumanity among us. There is no middle ground.

Of course, there will be vigorous debates over the exact definition of content that is so socially destructive that it deserves to be forbidden, hidden, or deprioritized in searches and newsfeeds. Overt racism, defamation, and denial of basic scientific facts are good places to start, particularly when they directly endanger lives and livelihoods. Review boards are critical to that definitional process.[14]

Ideally, they will reflect platform users' values and domain experts' considered positions, and will include attorneys capable of difficult balancing between free expression, privacy, and other values.

Even more important is a rethink of the relationship between new media and old. Most journalists, for all their well-documented faults, could be relied upon to at least try to verify the accuracy of news, thanks in part to legal requirements; a reckless lie can provoke a crippling defamation lawsuit. By contrast, lobbyists at technology firms have convinced all too many legislators and regulators to view their platforms as mere conduits for others, like a telephone line. This convenient self-characterization lets them escape defamation liability in numerous contexts.[15]

That impunity may have been understandable when Facebook and Google were fledgling firms. It is now anachronistic. Authoritarian propaganda has flooded news feeds and search results.[16] Facebook may disclaim responsibility for the spread of stories falsely claiming that the Pope endorsed Donald Trump or that Hillary Clinton is a Satanist.[17] Yet it has only taken on more responsibility for running political campaigns since then. As Alexis Madrigal and Ian Bogost have reported, "The company encourages advertisers to hand over the reins entirely, letting Facebook allot spending among ads, audiences, schedules, and budgets."[18] It also experiments with the content and appearance of ads. At some point, the AI is less finding audiences than creating them, less the conduit of a message than a coauthor of the message itself.

In many contexts, major technology firms are as much publishers as platforms, as much media as intermediary. Their own design choices mean that summaries of stories shared on Facebook, as well as those presented by Google's desktop and mobile displays, all tend to look similarly authoritative, whatever the source.[19] Thus, a story from the fabricated "Denver Guardian," falsely tying Hillary Clinton to an FBI agent's suicide, can appear as authoritative as a Pulitzer Prize–winning investigation debunking such conspiracy theories.[20] More directly, Facebook profits from fake news; the more a story is shared (whatever its merits), the more ad revenue it brings in.[21] Toward the end of the 2016 election, fake news outperformed real news in Facebook's ecosystem.[22] We also now know that Facebook directly helped the Trump campaign target its voter-suppression efforts at African Americans.[23]

Ethics was an afterthought, as long as the ad money kept rolling in. Safety and security officials at the firm were ignored, even after they told top managers that state and non-state actors were manipulating the platform in malign ways.[24] Of course, Facebook was not alone here; the recklessness of fringe news sources (and fecklessness of established media institutions) have also accelerated the rise of authoritarian and xenophobic leaders. But tech behemoths are scarcely innocent bystanders, either. Facebook has responded sluggishly to fake or misleading viral content.[25] When it takes Jack Dorsey of Twitter six years to pull the plug on a Sandy Hook truther like Alex Jones (who aggressively backed conspiracists who tormented the parents of slaughtered children, claiming that the parents had made the whole massacre up), something is deeply wrong online. The core problem is a blind faith in the AI behind algorithmic feeds and a resulting evasion of responsibility by the tech leaders who could bring bad actors to heel.

Tech behemoths can no longer credibly describe themselves as mere platforms for others' content, especially when they are profiting from micro-targeted ads premised on bringing that content to the people it will affect most.[26] They must take editorial responsibility, and that means bringing in far more journalists and fact checkers to confront the problems that algorithms have failed to address.[27]Apologists for big tech firms claim that this type of responsibility is impossible (or unwise) for only a few firms to take on. They argue that the volume of shared content is simply too high to be managed by any individual or team of individuals. There is a petulant insistence that coded algorithms, rather than human expertise, are the ideal way to solve problems on the platform. This argument ignores the reality of continual algorithmic and manual adjustment of news feeds at firms like Facebook.[28] Any enduring solution to the problem will require cooperation between journalists and coders.

BIG TECH'S ABSENTEE OWNERSHIP PROBLEM

After every major scandal, big tech firms apologize and promise to do better. Sometimes they even invest in more content moderation or ban the worst sources of misinformation and harassment. Yet every step toward safety and responsibility on platforms is in danger of being reversed thanks to a combination of concern fatigue, negligence, and the profit potential in arresting, graphic content. We have seen the cycle with Google and anti-Semitism. While the search giant at least labeled some troubling Nazi and white supremacist content in 2004, it backslid later, until it was called out by *Guardian* journalist Carole Cadwalladr and the prominent scholar Safiya Umoja Noble in 2016.[29] By 2018, there was again widespread concern that an algorithmic rabbit hole was AI-optimized to lure the unsuspecting into alt-right content, and worse.[30] As long as AIs drive an automated public sphere optimized for profits, expect the same pattern: flurries of concern spurred by media shaming, followed once again by irresponsibility.

Although such backsliding is not inevitable, it is a clear and present danger of AI driven primarily by the demands of shareholders of massive firms. Only scale can drive the profits investors seek, and only AI can manage that scale at its current level. As Evan Osnos observed in a revealing profile of Mark Zuckerberg, "Between scale and safety, he chose scale."[31] Children, dissidents, content moderators, hacked account holders, and other victims bear the brunt of that decision daily, as they endure the predictable externalities of social network gigantism. While behemoth banks have grown too big to fail, massive technology firms are simply too big to care.

The television was once known as the electronic babysitter, entertaining children whose parents were too busy or disadvantaged to provide them with more enriching activities. YouTube serves up millions of variations of cartoons to help a parent or babysitter find, with clinical precision, exactly what will best mesmerize a screaming toddler or bored child. But its vast stores of videos, all instantly accessible, make the job of parental supervision much harder. When some sick person(s) interspliced cartoon content with bland instructions on how to commit suicide, YouTube's AI, blind to meaning, ranked it just as highly as innocuous children's programming.

YouTube's ordinary content review processes can in principle improve over time to catch these videos more quickly. Some algorithmically generated content, however, raises another level of concern. Given the rise of automated remixes online, innocent cartoon characters like Peppa Pig may be chatting with friends in one video, then brandishing knives and guns in a "satire" auto-played right after. As artist James Bridle has observed, this goes beyond ordinary anxieties about decency. What is worrying about the Peppa videos, Bridle argues, "is how the obvious parodies and even the shadier knock-offs interact with the legions of algorithmic content producers until it is completely impossible to know what is going on"—be it artistic creativity, satire, cruelty, or perverse efforts to introduce small children to violent or sexualized content.[32] A context of no-context commingles all these forms, fraternizing incompatibles. Abuse predictably follows, raising deep questions about the role of AI in curation:

> At stake on YouTube is very young children, effectively from birth, being deliberately targeted with content that will traumatise and disturb them, via networks that are extremely vulnerable to exactly this form of abuse. It's not about intention, but about a kind of violence inherent in the combination of digital systems and capitalist incentives.

> The system is complicit in the abuse, and YouTube and Google are complicit in that system.[33]

These remix videos fuel a cynical game of human subjects research on children, where the chief marker of success is what brings in more ad revenue.

An ordinary television station broadcasting such dreck would bear some real consequences via viewership declines or sponsorship boycotts. But when journalists and the mass public think of "the algorithm" as some impossible-to-control Frankenstein's monster, YouTube can call itself a mere platform and blame an ever-shifting crowd of content creators, programmers, and search-engine optimizers for anything that goes wrong. Its cultivated irresponsibility previews how tech firms may take over more and more of our experience, taking credit for what goes right and deflecting blame for disasters.

In this scenario, AI serves as a mask or excuse, some ultra-complicated way of creating and ordering content that even its owners say they cannot control. That is an obvious dodge of responsibility. Live broadcasters will frequently delay their feed to allow for intervention if something goes awry. MySpace, an early social network, had a person review every picture before it was posted in Brazil, in order to stop the circulation of images of child abuse. Both YouTube and Facebook employ "cleaners" to review, post hoc, images and videos complained about by users or spotted via algorithms. For certain vulnerable populations (like children) or in sensitive times (for example, in the aftermath of a mass shooting motivated by racist hatred), they can do far more to invest in safety.

Despite persistent PR and pledges to "do better," there is a constant stream of news stories exposing how poorly the firms are performing in these roles. The *New York Times* found that YouTube was sometimes automatically recommending home videos of children in bathing suits after sexually themed content, reflecting pedophilic tendencies in its AI.[34] Reveal News found Facebook targeting persons addicted to gambling with ever more alluring video poker ads. At this point, if such stories declined significantly, it would be hard to know whether that actually signified public-spirited behavior by platforms, or merely the decimation of reporting capacities due to digital middlemen's diversion of ad revenues.

RESTORING RESPONSIBILITY IN ONLINE MEDIA

If Google and Facebook had clear and publicly acknowledged ideological agendas, adult users could grasp them and inoculate themselves accordingly, with skepticism toward self-serving content. The platforms' AI is better understood, however, as a powerful tool easily manipulated to the advantage of search-engine optimizers, well-organized extremists, and other fringe characters. A Google search for "Hillary Clinton's health" in September 2016 (after she briefly fainted) would have led to multiple misleading videos and articles groundlessly proclaiming that she had Parkinson's disease. In the fog of war that is modern political campaigning, the very fact of such a proliferation of defamatory videos (whatever their veracity) can become a major topic of public discussion. Manipulators know that simply raising an endless stream of spurious questions, no matter how poorly vetted or fabricated, can derail a candidate's chance to set the public agenda.[35]

Without any human editors or managers to take direct responsibility for algorithmic choices, online platforms can help the worst elements in societies appear credible and authoritative. For example, Google search results fed the racism of Dylann Roof, who murdered nine people in a historically black South Carolina church in 2015. Roof said that when he googled "black on white crime," he found posts from white supremacist organizations alleging a "white genocide" in progress. "I have never been the same since that day," he said.[36] #Pizzagate conspiracy theory spurred at least one gunman to "investigate" baseless allegations of sexual abuse at a Washington pizzeria while toting an AR-15. In the fever swamp of automated search results, support for climate denialists, misogynists, ethno-nationalists, and terrorists easily grows and spreads.[37]

Google's autocompletes—its automatic effort to anticipate the rest of a search query from its first word or two—have also sparked controversy.[38] They often repeat and reinforce racist and sexist stereotypes.[39] Google image search absurdly and insultingly tagged some photos of black people as gorillas. Within a year, Google "solved" the problem by turning off the gorilla classifier for all photos; nothing, not even actual gorillas, would be tagged as gorilla thenceforth. That quick fix does little or nothing to solve the underlying problems with the racist and sexist data now being used to train machine-learning systems or to reduce their vulnerability to attack.[40]

Many elites encountering the types of racist or sexist stereotypes documented in Google image results may simply take them as a tragic reflection of Google users' benighted attitudes. The users supply the data, after all. This shift in responsibility is not convincing. As Safiya Umoja Noble points out in her book *Algorithms of Oppression,* drawing on her expertise both as an academic and as a businesswoman, when the price is right, digital giants are willing to intervene and change results (for example, in commercial contexts when selling ads). They need to be just as accountable when the public interest is threatened.

Noble has also argued for slowing down new media. Instant uploading and sharing makes it "impossible for the platform to look at all of it and determine whether it should be there or not," Noble notes.[41] Cyberlibertarians argue that this simply means that human oversight is obsolete. Not so, counters Noble. These are human-serving systems that can be redesigned for human-serving ends. News stories about certain topics may need some vetting; delay can allow that to happen. It is not as if every one of millions of posts needs to be reviewed; rather, once a story has been "cleared," it could then be shared instantly. This deceleration of virality would also allow for more thoughtful decisions about limiting (or, in some cases, increasing) "algorithmic amplification." This amplification can lead to self-reinforcing feedback loops totally unrelated to the quality or usefulness of the underlying information.

Some tech apologists will profess horror at a proposal like this, calling it censorship. However, there is no magic free speech "default" on offer here. The platforms have made choices about whether to allow users to direct-message anyone (or only those who friend / follow them), how long hosted videos may be, and how rapidly stories spread or are suggested to other users. At present, those decisions are driven almost entirely by the profit motive: what communicative arrangement best maximizes advertising revenue and user engagement. These calculations can be made instantaneously and will often limit (and sometimes severely limit) the dissemination of a post. All Noble is suggesting is that more values—including those that take more time to discuss and apply—enter into the calculus. Like ensuring there is a "human in the loop" of otherwise autonomous weapon systems, Noble's "slow media" approach is wise.[42] It not merely circumvents unintended consequences. It also forces so-called mere platforms to take on the responsibilities of the media role they are now playing.

There are also concrete examples of ambitious curation projects arising, to provide higher quality (if less rapid and comprehensive) perspectives on the news, or key topics of interest to readers. For example, technology critic Evgeny Morozov leads a team of analysts who publish The Syllabus, a web-based collection of carefully chosen media in dozens of interest areas (from health to culture, politics to business). Echoing the commitment to complementarity expressed in the first new law of robotics, they promise a "novel and eclectic method that pairs humans and algorithms to discover only the best and the most relevant information."[43] A cynic may scoff at The Syllabus's aspiration to organize vast sets of information more effectively than the wizards of Silicon Valley. Yet it is hard to compare Morozov's team's remarkable series of media on the "politics of COVID" with the emergent "infodemic" of misinformation available on platforms, and simply assume that Facebook's, Twitter's, and Google's tens of thousands of employees are doing a much better job than Morozov's small crew.[44]

Whatever the substantive merits of The Syllabus's idealistic mission, all should acknowledge that a world of hundreds or thousands of trusted recommenders like it would decentralize power far more effectively than incremental reforms to Facebook or Google can. Vortices of online attention, they exercise disproportionate influence on what we watch and read. As the implications of this concentrated power to persuade become clearer, more competition law experts are calling for their breakup.[45] For example, while Mark Zuckerberg may argue that his company's control of Instagram, Facebook, and WhatsApp enables precisely targeted, relevant ads, it also provides a fat target for psyops firms and state actors seeking to manipulate elections. Even when such bad actors are controlled, it is illegitimate for one firm to have that much data from (and power over) users. Dividing up these essential communicative facilities would reduce the stakes of now-momentous corporate policies. It would also make it easier for media firms to demand more compensation for the human expertise they provide, rebalancing a digital advertising landscape now tilted in favor of tech.[46]

THE WRECKING BALL OF ONLINE ATTENTION

Complaints about the political and cultural influence of online intermediaries are of long standing. Their commercial effects are also coming under fire. A one-size-fits-all internet portal can't serve everyone equally well—especially when one goal (profitability) supersedes all others. Vulnerable, easily exploited populations can easily become profit centers.

Consider how Google's search engine has become a one-stop shop for all manner of information, even for those in desperate circumstances. For example, when those addicted to drugs seek help, they often turn first to a Google search, typing in terms like "rehab center" or "opioid addiction" in the firm's iconic search box. They have little idea of the swirling ecosystem of data brokers and hucksters in search of large payments for cheap, unvetted treatment services. For some time, a search for "rehabs near me" would generate ads for third-parties who made money by selling leads. These lead generators often did little to validate whether those paying them were legitimate addiction recovery services. Sometimes they would run hotlines that claimed to be unbiased but actually received money from the companies they referred clients to. Journalist Cat Ferguson concluded that the resulting chaos "pushed many people away from getting the help they needed—and, in extreme cases, could look a lot like human trafficking."[47] A vicious feedback loop can develop: the firms that spend the most on marketing (and least on actual care) can crowd out the better, care-driven ones with a fusillade of paid ads, which gain more vulnerable clients thanks to the prominence they purchase, which can then fund more rounds of ads. It amounts to exactly the type of "arms race for attention" that the third new law of robotics is designed to prevent.

Although boosters for machine learning acknowledge that such manipulation is a real problem, they assert that there are cheap ways to address it without regulatory oversight. In order to eliminate commercial distortions, they say, invite volunteers to populate contested search results pages. For example, user-generated content on Google Maps can lead those seeking help to clinics that are closest to them. In their book *Machine, Platform, Crowd: Harnessing Our Digital Future*, Andrew McAfee and Erik Brynjolfsson frame this kind of interaction between a large internet company (the platform) and its users (the crowd) as a mostly productive synergy.[48] Large platforms like Google can use machine learning for many tasks, but then find ways to draw on the "crowd" of volunteer internet users to fill in gaps where more or better data is needed.

Concerned about potential negative publicity from its referrals to unsavory addiction clinics, Google doubled down on the "crowd" option. It stopped taking ad money from addiction treatment centers, leaving results to appear "organically." Scam artists, however, went on to manipulate the user-generated content sections of Google Maps. Anonymous saboteurs began to change the listed phone numbers of reputable addiction clinics to numbers for lead generators, who would then try to herd them toward substandard clinics.[49] Whatever the merits of the "crowd" in other contexts, when serious money is at stake, trust-based systems are easily abused.[50]

Fed up by the trickery of unscrupulous addiction centers, Florida has banned deceptive marketing for rehabilitation clinics.[51] It is unclear how effective the law has been. Nearly every state in the United States has some consumer protection law, and such laws probably already banned the most troubling practices of addiction clinic lead generators, as well as the unscrupulous addiction clinics themselves, before they came to public attention. Law enforcement officials just don't have the resources to investigate every suspicious occurrence, especially when digital manipulators can so easily hide their identities and set up shop in new places once an investigation begins. So large internet firms must do some of the policing themselves, lest vulnerable groups be overrun by predators trying to take advantage of them.[52]

JUST AND INCLUSIVE GOVERNANCE IS NOT ALGORITHMIC

When the US Congress called on Mark Zuckerberg to testify about his social network's repeated failures to police fake news, voter manipulation, and data misuse, the Facebook CEO had a common refrain: we're working on artificial intelligence to solve it. Critics are skeptical. As one puckish headline writer put it, "AI will solve Facebook's most vexing problems. Just don't ask when or how."[53] In an earnings call, Zuckerberg conceded that progress might be slow on some fronts: "It's much easier to build an AI system to detect a nipple than it is to detect hate speech," he explained. In other words, a self-identity as an "AI company" made the prohibition of a form of nudity on the platform a higher priority for its managers than cracking down on hate speech.

That throwaway line glossed over major controversies about censorship, values, and responsibility in online spaces. There are widely divergent views on whether nudity should be banned on a massive social network, and which should be a higher priority. Nor are site policies consistent or implemented automatically. One of Facebook's properties, Instagram, does not permit users to post images of naked bodies or digital representations of them, but it does accept painted or drawn nude portraits.[54] One plausible rationale for the distinction is to protect artistic depictions of nudity while barring pornographic or exploitative ones. Jurisprudence on obscenity revolves around such distinctions; however, they are also embedded in a much broader framework of evaluation. Without such context, the physical / digital distinction is not a good proxy for the aesthetic / non-aesthetic divide: one can easily imagine drawn pornography and digital drawings or photographs of nude bodies with great artistic merit.

The digital / non-digital divide is, however, very useful in another way: it is much easier to imagine a future machine-vision system effortlessly parsing the difference between paint and pixels than convincingly distinguishing between a still from the beginning of a pornographic film and a photograph documenting the lives of sex workers in a tasteful and artistic way. The former is a mere pattern-recognition task, and we can imagine hundreds of thousands of images uncontroversially classified as one or the other, perhaps by workers on Amazon's Mechanical Turk platform (and perhaps even via unsupervised machine learning).[55] The latter is a matter of moral and aesthetic judgment—and that judgment may change over time, or vary by place. We should not allow the limits of a tool (AI) to dictate a partial or ham-handed approach to a difficult cultural and political question (how much nudity is appropriate in social media venues easily accessible to children).[56]

In well-ordered polities, communities debate the scope of the obscene, regulators must decide what to suppress and what to ignore, and courts can step in to defend the rights of those unfairly singled out by prejudiced or biased decision makers. Even if we could, in principle, hard-code judgments in 2020 about what types of images were acceptable on their platforms and which were not, of what value would that be in 2030 or 2040? Views change, and the audiences and purposes of platforms expand and contract, as Tumblr learned to its chagrin after its pornography ban. Governance is an ongoing obligation here, not something that can simply be mechanized once for all time. Nor is volunteer user input an easy solution here, as Amazon learned when homophobes labeled thousands of books about gay culture "adults only" in order to trigger automated steps to hide that content. Motivated extremists, aided by bots and hackers, can easily overwhelm solid majorities. The so-called democratizing tendencies of the internet have run into a buzz saw of absentee monopolists, motivated manipulators, and automated "speech" via bots.

Hard-core technophiles may claim that a decision about whether nudity is aesthetic or prurient, deserving or undeserving of attention, is merely a matter of taste. They tend to deem it something "beneath" machine intelligence, a form of knowledge that is second-rate because it cannot be reduced to an algorithm. One can just as easily subvert that hierarchy, valuing non-standardizable judgments above those that are data driven or rule enabled. We may not want to invest the resources necessary to make these decisions tractable to human judgment or contestation. But that is a discretionary decision about resource allocation, not the dictate of inevitable technological progress toward artificial intelligence.

THE RIGHT TO BE FORGOTTEN: A TEST RUN
FOR HUMANE AUTOMATION

Whenever those hurt by online manipulation propose regulation, a predictable chorus responds that however badly tech firms are doing, government intervention would be worse. Yet we already have a great case study in successful, humane automation in the public sphere: a branch of European regulation of automated name-search queries known as the Right to be Forgotten (RtbF). To understand the importance of that right, imagine if someone built a website about the worst thing that ever happened to you or the worst thing you ever did, which went on be in the top search results of your name for the rest of your life. In the United States, you have no legal recourse; corporate lobbyists and libertarian ideologues have united to make regulating such results politically radioactive. In Europe, however, the RtbF ensures that the remorseless logic of purely algorithmic search results does not tag persons with a permanent digital "scarlet letter."

The RtbF is a good test run for governing technology with the values of a broad political community (and not just the values of those who happen to own the tech). Before the RtbF emerged, Google's leaders repeatedly insisted on automating as much of a search as possible, consistent with *searchers'* experience of quality.[57] The company refused to take into account the interests of those *searched*. That's a classic strategy in AI optimization—dropping one whole set of considerations in order to focus on simpler desiderata (like ad revenue), which a machine can parse. Yet just as governments should not neglect or ignore whole sets of people in order to make jobs easier for legislators and bureaucrats, those behind AI need to take on the full complexity and burden of their effects in the world.[58] These firms are effectively governing digital reputations, and they need to act accordingly.

Citizens of the European Union can remove from search results information that is "inadequate, irrelevant, no longer relevant or excessive," unless there is a greater public interest in being able to find the information via a search on the name of the data subject.[59] That's a standard that is at least as hard to parse as the difference between art and pornography. But web users, search engines, and European officials have been working hard to strike a balance between individual interests in privacy and the public's right to know.

For example, consider the story of a woman whose husband was murdered in the 1990s and who then found—decades later—that whenever anyone googled her, the story of the murder was in the results. This is a classic example of a potentially excessive impact from one piece of data on a person's life. It is not false, so defamation law cannot help the woman. But the RtbF can. The woman petitioned to have these stories removed from search results about her. A human at Google complied, eliminating the results. There was no countervailing public interest in the public having this story at the ready whenever they googled the widow's name.[60]

When the Court of Justice of the European Union announced the right to be forgotten, free speech activists (including many funded by Google) decried it as an attack on freedom of expression. These complaints were overblown; the RtbF does nothing to affect the source of the information at issue, and there is no free speech right to ensure that negative information in a private database is what people see attached to a person's name. Moreover, Google and public authorities have been judicious about approving these requests. The legal scholar Julia Powles has compiled some key RtbF decisions, contrasting successful and rejected delisting requests.[61] Politicians with fraud convictions are routinely denied the right to remove news about those misdeeds from their search results. Patients whose HIV status has been exposed have removed links to stories publicizing it. Powles's excellent work demonstrates how careful application of legal and ethical principles can lead to nuanced, contextual judgments that do reputational justice to victims of unfairly salient results while respecting the public's right to easy access to important facts about public figures or those otherwise in a position of trust.

Battles over the RtbF touch on many sensitive matters. Child molesters in Europe have also lost their bids, even when their crimes were decades old; the public's right to know outweighed their right to a fresh start. The Japanese Supreme Court embraced the same rationale when it overturned a lower court's decision to grant a sex offender's RtbF request.[62] As the Japanese case suggests, even experienced jurists may disagree about the scope and force of a right like the RtbF. For those seeking algorithmic justice, delivered as predictably and implacably as numerical calculation, that is a defect of human judgment. But when we consider how subtle the application of power should be in new digital domains, human flexibility and creativity is an advantage.[63] We should think hard about how long we stigmatize persons and how technology can unintentionally multiply the effective term of a shaming. Experts in law, ethics, culture, and technology must all be part of this conversation.[64]

Many RtbF cases involve difficult judgment calls: exactly what human workers exist to evaluate and adjudicate. Content moderators already eliminate pornographic, violent, or disturbing images, albeit at low wages and in often inhumane conditions.[65] Given the extraordinary profitability of dominant tech firms, they can well afford to treat those front-line workers better. One first step is to treat content moderation as a professional occupation, with its own independent standards of integrity and workplace protections.

Search-result pages are not a pristine reflection of some preexisting digital reality, as victims of misfortune, bad judgment, or angry mobs know all too well. They are dynamic, influenced by search-engine optimizers, engineers within Google, paid ads, human reviewers of proposed algorithm changes, and many other factors. No one should assume that these results necessarily amount to an expression of truth, a human opinion, some company stance, or some other aspect of expression that merits robust protections for freedom of expression. If the aims of privacy, antidiscrimination, and fair data practices are to be realized in a digital age, search engines' status as data processors and controllers—their own dominant self-characterization—must take precedence over the "media defendant" status they opportunistically invoke. Algorithmic arrangements of information should be subject to contestation based on societal standards of fairness and accuracy. The alternative is to privilege rapid and automatic machine communication over human values, democratic will formation, and due process.

TAMING THE AUTOMATED PUBLIC SPHERE

The right to be forgotten stands as a proof of concept: even the most comprehensive computational processes can accommodate human values. Other legal initiatives would help counter the discrimination, bias, and propaganda now too often polluting (and even overwhelming) online spaces. Governments can require platforms to label, monitor and explain hate-driven search results, and other deeply offensive content.

Lawmakers should, for instance, require Google and other large intermediaries not to link to sites denying the existence of the Holocaust, or at least to severely down-rank their prominence. There are some ideologies that are so discredited and evil that they do not deserve algorithmically generated publicity. At the very least, governments should require educational labeling in the case of obvious hate speech boosted by manipulative methods. To avoid mainstreaming extremism, labels may link to accounts of the history and purpose of groups with misleadingly innocuous names which actually preach discredited and dangerous ideologies. Without such labels, an obsessive band of bigots with basic coding skills can hack the attention economy, taking advantage of failures of automated systems to properly label and explain information sources.[66]

Automated bots have done much to promote extremism online, and governments must also rein them in. Massive tech platforms fight only a half-hearted battle against such accounts. These platforms have dual loyalties. They know that users don't care for annoying replies from, say, @rekt6575748, or video view counts inflated by automated click farms. On the other hand, they do not have much competition, so there is little reason to fear user defection. Meanwhile, bots inflate platforms' engagement numbers, the holy grail for digital marketers.

In 2012, YouTube "set a company-wide objective to reach one billion hours of viewing a day, and rewrote its recommendation engine to maximize for that goal."[67] "The billion hours of daily watch time gave our tech people a North Star," said its CEO, Susan Wojcicki. Unfortunately for YouTube users, that single-minded fixation on metrics also empowered bad actors to manipulate recommendations and drive traffic to dangerous misinformation, as discussed above. To help identify and deter such manipulation, both social networks and search engines should crack down on manipulative bots. If they fail to do so, laws should require every account to disclose whether it is operated by a person or a machine. All users should be able to block bot accounts as part of their user settings—or better, such settings should default to blocking bots, with affirmative action required to open a user's feed to their automated emissions. Some technophiles worry that such restrictions could interfere with free expression. But a purely automated source of information and manipulation, with no humans directly taking responsibility for it, should not enjoy such rights. Governments can ban sound-emitting mini-drones from invading our personal space to share news, views, or advertisements, and the same logic applies online.

Free speech protections are for people, and only secondarily (if at all) for software, algorithms, and artificial intelligence.[68] Free speech for people is a particularly pressing goal given ongoing investigations into manipulation of public spheres around the world. As James Grimmelmann has warned with respect to robotic copyright, First Amendment protection for the products of AI could systematically favor machine over human speech.[69] Whereas human speakers must worry about defamation law or other forms of accountability if they lie or get things wrong, how can an autonomous bot be held responsible? It has no assets or reputation to lose, so it is impossible to deter it.

It is much easier to mimic popular support than to generate it.[70] When voters and even governments are unable to distinguish between real and counterfeit expression, what Jurgen Habermas calls "democratic will formation" becomes impossible. As a growing body of empirical research on the troubling effects of bot "expression" shows, in too many scenarios, bot interventions are less speech than anti-speech, calculated efforts to disrupt deliberation and fool the unwary.[71] To restore public confidence in democratic processes, governments should require rapid disclosure of the data used to generate algorithmic speech, the algorithms employed, and the targeting of that speech. The fourth new law of robotics (requiring attribution of AI to its inventor or controller) demands such transparency. Affected firms may assert that their algorithms are too complex to disclose. If so, authorities should have the power to ban the targeting and arrangement of information at issue, because protected speech must bear some recognizable relation to human cognition. Without such a rule, a proliferating army of bots threatens to overwhelm actual persons' expression.[72]

Authorities should also consider banning certain types of manipulation. The UK Code of Broadcast Advertising states that "audiovisual commercial communications shall not use subliminal techniques."[73] There is a long line of US Federal Trade Commission guidance forbidding misleading advertisements and false or missing indication of sponsorship. States will also need to develop more specific laws to govern an increasingly automated public sphere. California recently required digital bots to at least identify themselves on social media sites.[74] Another proposed bill would "prohibit an operator of a social media Internet Web site from engaging in the sale of advertising with a computer software account or user that performs an automated task, and that is not verified by the operator as being controlled by a natural person."[75] These are powerful, concrete laws to assure that critical forums for human communication and interaction are not overwhelmed by a posthuman swarm of spam, propaganda, and distraction.

Whatever the science fictional appeal of speaking robots, no one should romanticize manipulative bot-speech. The logical endpoint of laissez-faire in an automated public sphere is a continual battle for mindshare by various robot armies, with the likely winner being the firms with the most funds. They will micro-target populations with "whatever works" to mobilize them (be it truths, half-truths, or manipulative lies), fragmenting the public sphere into millions of personalized, private ones. It does not represent a triumph of classic values of free expression (autonomy and democratic self-rule); indeed, it portends their evaporation into the manufactured consent of a phantom public. Undisclosed bots counterfeit the basic reputational currency that people earn by virtue of their embodiment, blatantly violating the second new law of robotics.

THE ULTIMATE ROBOCALLER

This counterfeiting problem may soon escape the digital world and enter the real. When Google Assistant debuted in 2018, it wowed audiences, particularly when CEO Sundar Pichai demonstrated Duplex, an AI designed to make appointments. "We want to connect users to businesses, in a good way," Pichai explained. "Sixty percent of small businesses don't have an online booking system set up," and many customers find calls annoying.[76] As he spoke, a screen showed the typing of a command: "Make me a haircut appointment on Tuesday morning anytime between 10 and 12." At that command, Duplex instantly made a call to a salon and interacted with the person who answered the phone. "Hi, I'm calling to book a women's haircut for a client," Duplex said. It was unclear at the time whether the "voice" of Duplex was a synthesized voice or that of an actual person. Pichai played clips of Duplex asking for a restaurant reservation. The bot was extraordinarily lifelike. When it voiced "mmmhmm" in response to a salon employee's comment, the crowd oohed and ahhed with delight. In a more complex exchange, Duplex processed a restaurant worker's advice that a reservation wasn't really needed certain nights. This was much more than a phone tree. Google seemed to have mastered many basic tasks of communication, drawing on a massive database of phone conversations.

Making a restaurant reservation may not seem like a massive achievement. The project leads behind Duplex were quick to remind the starstruck that the AI had very narrow conversational abilities.[77] Yet it is not hard to extrapolate the road Duplex puts AI research on. Consider, for example, the millions of messages in the databases of groups such as Crisis Text Line, which texts clients through problems ranging from calamities to garden-variety depression.[78] It is expensive to train persons to respond to clients. Could we, one day, imagine an AI-driven resource that matches up complaints, worries, and expressions of despair to the language that comforted such distress in the past? Will Duplex technology be the future of suicide hotlines? Less dramatically, does it presage the extinction of the customer service representative, as every imaginable scenario is chirpily covered by bots parsing a sprawling database of complaints and responses?

As of now, the answer appears to be no. A tidal wave of online outrage soon overwhelmed Google. Deceiving unwitting employees into thinking they are dealing with a human being is a technical advance, but not a moral one. A recording, no matter how skilled and personalized, makes a kind of claim on our time and attention far different than a live person on the phone. After a series of bruising articles in the press, Google grudgingly conceded that Duplex should identify itself as an AI when it makes calls.[79] The company also set up an "opt-out" for businesses that did not want to receive its calls—adapting its service to extant US Federal Communications Commission requirements for robocallers. Ironically, Duplex itself may be getting lost in the shuffle of spam calls, as restaurant employees ignore calls without an identifiable name attached to them.

The backlash to Duplex was at least as remarkable as the product itself. In the past, the main complaint about AI and robotics was their rigidity. The merely mechanical rankled because it was so far from the spontaneity, creativity, and vulnerability of the human. More recently, however, human-computer interaction scholars theorized the problem with the "almost-but-not-quite-human" robot.[80] Many in affective computing—a field that tries to simulate human emotion in robots—believe that with enough data, sophisticated enough audio simulations of voices, and more lifelike faces, their creations could finally escape the uncanny valley. But the backlash to Duplex suggests another possibility. "Google's experiments do appear to have been designed to deceive," said Oxford researcher Thomas King when interviewed by noted tech journalist Natasha Lomas.[81] Travis Korte commented that "we should make AI sound different from humans for the same reason we put a smelly additive in normally odorless natural gas."[82] Korte's comment puts a new spin on the old saw that "big data is the new oil"—you need to know when you're dealing with it, lest it blow up in your face. It's an entirely different matter to have a conversation with a human (however brief) and to be subject to the manipulation of a human-mimicking AI controlled by one of the largest companies on earth.

Duplex may seem like an outlier, a deputation of small talk to AI with little lasting impact. However, it is the tip of an iceberg of AI-driven intermediaries with increasing power over our communicative landscape. What Duplex does for conversational snippets, YouTube does for videos and Facebook for media generally: find an "optimal" match given a pattern of words. We don't often think of AI when we scroll or search online, and that is a sign of its commercial success. Firms have so much data about our viewing habits, our digital journeys across the web, our emails and texts and location, that they are building compelling, even addictive, infotainment machines.[83] The less we think about how we are being influenced, the more powerful influencers can become.[84]

A RETURN TO JOURNALISTIC PROFESSIONALISM

So far, governments are only beginning to wake up to the threat posed by the automated public sphere. Germany has led the way, with legislation imposing clear responsibility on platforms for fake news. Inadequacies of AI detection systems are no longer an excuse there. The Netherlands has also empowered officials to punish platforms that feature hate speech or lies. Still, media regulators in both countries face criticism from a press that perceives any restrictions—even on the platforms that are now draining their revenue away—as an assault on hard-won independence. Matters are even worse in the United States, where a prevaricator-in-chief managed to use "fake news" as an epithet against the very outlets trying to expose his misdeeds. Politicians and celebrities have harassed platforms like Facebook for the minimal steps it has taken to reduce the visibility of some conspiracy theories and hate speech. This pressure means that legislation to improve the online media landscape will face an uphill battle.

Even if governments fail to require auditing and attribution for expression in the automated public sphere, large internet platforms can take steps toward self-regulation (if only to assuage skittish advertisers). They can hire more journalists to vet stories once they go viral and respect their input by adjusting algorithms to limit the dissemination of hate speech and defamatory content. Facebook has begun cooperating with non-profit organizations on this front, but the effort is poorly funded and limited in impact.[85] Mere volunteers cannot tame the tide of weaponized information—and even if they could, it is not fair to expect them to work for free. Indeed, to the extent these fact-checking initiatives work, there will be great opportunities for the ultimate funders of fact checkers to impose their own agendas. Better for tech firms to recognize and value the professional prerogatives and identity of journalists and editors and employ them as equal partners with coders and engineers as they construct the digital public sphere of the future.

This is not yet a popular position. Many in Silicon Valley believe that crafting newsfeeds is an inherently algorithmic function, to be supervised by the engineers.[86] But Facebook itself thought differently for a while, having hired human editors for the "trending topics" box on search feeds. Admittedly, these were low-status contract workers, who were unceremoniously dumped when a thinly sourced news story asserted that conservative content was being suppressed.[87] Shortly thereafter, Facebook was swamped by trending fake news, which gave extraordinary publicity, exposure, and moneymaking opportunities to charlatans and liars. The real lesson here is that human editors at Facebook should be given more authority, not less, and that their deliberations should be open to some forms of scrutiny and accountability. These jobs are ideal for professional journalists, to be treated with the same level of respect and autonomy that large tech firms grant those in engineering roles. Journalism schools are already teaching skills in coding, statistics, and data-driven analysis to writers capable of bridging the divide between the tech and media sectors. There are also many local professional journalists available for such roles. After decades of layoffs, the US had roughly eighty-eight thousand newsroom jobs at the beginning of 2020; a few months later, thirty-six thousand of those were eliminated.[88] A natural economic transition would bring some of this expertise to bear on the automated public sphere.

Some communication scholars have resisted the idea of professionalization of online content creation, curation, and delivery in the name of citizen journalism that would democratize the power of the press to anyone with a computer and an internet connection. While a beautiful ideal in theory, in practice it has faltered. A failure among the de facto sovereigns of the internet to distinguish between stories on the real *Guardian* and the "Denver Guardian" is not simply a neutral decision to level the informational playing field. Rather, it predictably acceler-ates propaganda tactics honed by millions of dollars of investment in data, PR, and spin-doctoring. Shadowy quasi-state actors are well prac-ticed in the dark arts of bias, disinformation, and influence.[89] Freedom for the pike is death for the minnows.

In our time, for better or worse, vast conglomerates like Facebook and Google effectively take on the role of global communication regu-lators. They must take responsibility for this new role or be broken up in order to make way for human-scale entities capable of doing so. The public sphere cannot be automated like an assembly line churning out toasters. The journalistic endeavor is inherently human; so, too, are edi-torial functions necessarily reflective of human values.[90] There will be deep and serious conflicts in this endeavor—over the proper balance between commercial interests and the public interest in assigning prominence to different sources, over how much transparency to give decisions made about such issues, and over how much control indi-vidual users should have over their newsfeeds. These are matters of utmost importance to the future of democracy. They can no longer be swept under the rug by managers more interested in stock returns and artificial intelligence advances than the basic democratic institutions and civil society that underwrite each.

HUMANIZING THE AI OF AUTOMATED MEDIA

Automated media has rapidly reorganized commercial and political life. Firms have deployed AI to make the types of decisions tradi-tionally made by managers at television networks or editors at news-papers—but with much more powerful effects. Platforms based in the United States and China have changed the reading and viewing habits of hundreds of millions of people around the world. Disruption has hit newspapers and journalists hard as internet platforms siphon away tra-ditional media revenues. And the transition to algorithmically cu-rated social media has been downright terrifying and tragic for some vulnerable groups—for example, minority Muslims subjected to mob violence whipped up on social media. Disastrous for many democracies, the alien intelligence of automated media has nevertheless emerged triumphant in the media marketplace.

I call this intelligence "alien," and not merely artificial, for several reasons. First, it is far removed from recognizable patterns of cognition among those it seeks to replace. For example, in choosing a story for the front page of a newspaper, editors primarily exercise news judgment. There are commercial considerations, to be sure, and perhaps some hidden motives, but the decision is in principle articulable and attributable to persons. The decision process for placing articles in a digitized newsfeed (or for many forms of algorithmic media optimized for digital environments) is kaleidoscopically fragmentary. Past user history, the reputation of a news source, efforts to boost or manipulate its popularity, and hundreds of other variables are instantaneously commensurated—combined in such a way as to crush down all manner of qualitative factors into numerical scores—and almost always blackboxed, immune from outside scrutiny. It is much harder to study than mass media, given the personalization of content to hundreds of millions of individuals, and the near-impossibility of obtaining a truly representative sample of who saw what, when.

Alien intelligence also suggests the distance of global technology company bureaucrats from the real world consequences of their algorithms. These managers may release a service in a country whose language is unknown not merely to them, but also to anyone several reporting levels down. They follow in the worst tradition of what Thorstein Veblen called "absentee ownership": distant controllers poorly understanding their business's context and impact.[91] When a massive firm buys a store thousands of miles away from its headquarters, it tends to assess its performance in crude terms, with little interest in the community in which the store is embedded. Once under new ownership, the store may neglect traditional functions it had previously served, in order to maximize revenue in accordance with its absentee owner's demands. In contrast, a present owner, resident in the community, is more likely to run the store in a way that comports with community interests and values, since the present owner will enjoy improvement (or suffer deterioration) wrought by the business in the community. We can say much the same for local journalism and media, which has been devastated by the same digital transition that has so enriched Facebook and Google. However badly it may have served its local audience, at least it had some real stake in the area. A newspaper like the *Baltimore Sun* cannot survive if its eponymous city collapses; tech giants will barely notice the tragedy.[92]

Finally, users of these automated media are all too prone to alienation, a dispiriting sense of meaninglessness and fragmentation, which can easily curdle into extremism or withdrawal. A proper analysis of this alienation goes well beyond the contextless critiques of "echo chambers," common in the twenty-first century's first wave of social media criticism. Rather, it should echo the critical theory penned in reaction against the authoritarianism of the 1930s. Particularly when it comes to politics, when citizens are thrown into what Rahel Jaeggi calls a "relation of relationlessness," indifferent or even menaced by all mainstream political parties, the stage is set for a radical degradation of the public sphere.[93] The most successful authoritarian populists of our time have exploited this widespread alienation, channeling the resulting discontent into destructive political programs.

The new laws of robotics cannot solve all these problems. These are deep challenges, not only to democracy, but to political community of all kinds. Nevertheless, renewal of commitments to complementarity, authenticity, cooperation, and attribution is essential to the future of humane media. Revitalizing journalism, editing, and curation as stable and well-compensated professions will at least give us a fighting chance against the tyranny of AI intermediaries (as well as the army of PR experts and consultants so skilled at manipulating them). Requiring bot disclosure will give users and platforms essential information they need to determine for themselves the mix of computational and human interaction they want. Rebalancing computational and professional judgment should reduce alienation as well, as would a cultural shift away from passive reliance on intermediaries for news and analysis, and toward more specialized curators with some more direct or local connection to their audience.

AI can promote free expression, *if* it is properly bounded by the four new laws of robotics. AI cannot replace journalistic judgment—or even many of the army of "content moderators" who now patrol for illegal, repulsive, or hate-filled posts. By provoking an arms race of targeted marketing among firms, politicians, and propagandists, major AI-driven firms have hijacked revenue streams that once supported journalism far more robustly. Some of that money should be directed back to professional journalists and to both old and new truth-producing (and consensus-shaping) institutions.[94] This is particularly important as makers of bots and deep fakes become ever more sophisticated. The fourth new law of robotics, requiring immediate and clear attribution, should guide regulation of the "fake media problem," as should the second law, barring the deceptive mimicry of human action. No one should need a degree in digital forensics to figure out whether tweets and videos represent authentic content or are merely a confected spectacle.

Platforms can be reformed, and communications scholars have developed inspiring visions of AI designed to complement, rather than prematurely replace, the editors, journalists, and creatives now being squeezed by powerful intermediaries. While meritorious, to reach fruition, these proposals will need the supportive political-economic environment provided by the four new laws of robotics. If properly implemented, these laws will also help prevent the worst excesses of the automated public sphere.

5

▪ Machines Judging Humans

Imagine being denied a job simply because of the pitch of your voice. Firms are now contracting with recruiters that use face and voice recognition to assess "nervousness, mood, and behavior patterns," all to judge who is a good "cultural fit" for a company. Companies as large as PWC, L'Oreal, Mars, and Citibank have all deployed the AI.[1] To those promoting AI as a human resources tool, face and voice scanning are just the next logical step from résumé-sorting software. If a machine can reject hundreds of résumés automatically, why not allow it to judge more ineffable matters?

Discrimination is a clear and present danger. The software may exclude minorities who are underrepresented in the AI's dataset. Feminists are quick to remind firms that the look and sound of their "model employees" may have as much to do with entrenched sexism as with genuine competence. If women were not part of the management team of the past, they cannot inform the dataset predicting future stars. Reinforcing these concerns, the advocacy group Upturn has published a comprehensive report on problematic hiring algorithms, recommending "active steps to detect and remove bias in their tools."[2] For example, employers could give second chances to particularly promising candidates knocked out by the filtering AI.

Face and voice parsing also undermines dignity and autonomy. There is a big difference between having one's résumé analyzed by an AI and being judged on mysterious aspects of one's demeanor and bearing. I choose what to put on a résumé; I cannot choose whether my eyes open a millimeter too wide when my interviewer says something surprising.

Yes, job interviews by human beings can be awkward and forced; skilled bluffers may well talk their way into unmerited positions, while other, better recruits fail. Still, the merely mechanical analysis of demeanor, bearing, and appearance seems disrespectful. We know that the top managers in a firm are not chosen this way. Why relegate other employees to such a dehumanizing process?

As more machines judge humans, at least four reactions are possible. Some will try to find ways to turn the system to their advantage, reverse engineering ineffable badges of aptitude, cultural fit, or whatever else an AI presumes to predict. Another group will strive to improve these systems' accuracy; a third, smaller one will try to make them fairer, either via technical methods or regulation. A fourth group will agitate to ban machine judgments altogether, to keep them out of certain contexts, or to require an appeal to a "human in the loop" of decision-making.

This chapter focuses on the last two groups—those who would try to mend, or end, evaluative AI, which presumes to measure persons' reliability, creditworthiness, criminality, or general "fit" for jobs or benefits. Machine judgments of people are a far cry from image recognition in medicine, where everyone agrees that improving technology serves humanity.[3] Finding a tumor in a lung is not like singling out a ne'er-do-well. There is real controversy over whether all-seeing computational eyes should dot our schools, streets, and hospitals or should be monitoring our every keystroke. The same can be said of new ways of assessing trust and creditworthiness, such as "fringe" data gathering by fintech (financial technology) startups.

A movement for fairness, accountability, and transparency in machine learning has proposed many ways of improving the AI now estimating human effort, attentiveness, trustworthiness, and worth.[4] There is also a good case for banning many of these technologies preemptively, and then only licensing them in specific, case-by-case instances. For example, schools in general should not be subject to Hikvision's gaze (mentioned in Chapter 3), but we can imagine it being useful at, say, a hospital that has faced repeated charges of abuse or neglect. Even there, though, the idea of a computational judgment of care quality or worker value should be resisted. Humans owe one another attention and explanation in human terms and cannot outsource such critical roles to opaque and unaccountable software.

THE APPEAL (AND PERIL) OF ROBOTIC JUDGMENTS

Imagine you receive a summons in the mail to appear in before a judge. Arriving in the courtroom, you find no one there. A kiosk takes your picture and displays a screen instructing you to take a seat with a large number nine printed on it. Its voice, which sounds like a blend of Siri and Alexa, states that your case will be the ninth heard that day. Once everyone is settled, an announcement booms, "All Rise." A curtain at the front of the room reveals another screen, with an image of a man in a robe. This judge's face is one of the millions of fake faces spawned by generative adversarial networks.[5] It is a composite derived from images of judges appearing in a sample of ten thousand films commonly used by AI researchers. He is the AI version of the judge from central casting.

A robot bailiff calls your case, and you step past a gated barrier to a chair and desk set out for defendants. The judge-avatar begins to speak: "You have been found guilty of a critical mass of infractions of the law. Your car has gone at least five miles per hour over the speed limit ten times over the past two years. You downloaded three films illegally last year. You smoked marijuana at a party. According to our algorithmic sentencing, optimal deterrence for this mix of offenses is forty points deducted from your credit score, a fine of 5 percent of your assets, and your agreement to install a home camera, algorithmically monitored, for the next six months, to ensure you do not violate the law again. If you wish to file an appeal, please enter your grounds for appeal in the Swift Justice app just downloaded onto your phone. If not, please exit the door you came in through, and the kiosk will have further instructions."

That scenario may seem ridiculously sci-fi—Franz Kafka meets Philip K. Dick.[6] And yet it reflects three current trends in social control. The first is ubiquitous surveillance—the ability to record and analyze every moment of a person's day. The second is a quest for smart cities, which could dynamically adjust penalties and rewards for forbidden and recommended behavior. Futurists envision outcome-based policing that ditches old-fashioned law for more flexible evaluations of neighborhood order. In this ideal of order without law, if a loud party draws enough complaints, it's illegal; if not, it can go on. The third is a quest to replace people with machines. While security guards and police draw salary and pensions, robots simply require electricity and software updates. The future of guard labor may well be swarms of drones, ready to immobilize any criminal or suspect.

Artificial intelligence and robotics in law enforcement and governance are, at present, largely driven by an economic narrative, about using resources more effectively. A mix of drones, CCTV cameras, and robots may be far less expensive than battalions of police officers. They may be more effective, too, and even less likely to injure suspects or wrongly target the innocent. But there is a doubleness, a sense of loss as well as gain, when government moves too far along a spectrum of control.[7] Government always stands as both a protector and potential threat, a source of aid and oppression. Police can protect peaceful protesters from a hostile crowd, but they can also brutally arrest and imprison them. Even more vexingly, the benefits and burden of state force are all too often dispensed arbitrarily and discriminatorily. Police disproportionately inflict violence on minorities, ranging from African Americans in the United States, to Uyhgurs in China, to residents of favelas in Brazil.

Communities committed to the rule of law have found ways of curbing this unfairness. Civilian review boards and courts can censure police overreach, ordering damages when civil rights are violated. The Fourth Amendment of the US Constitution bars "general warrants," which are blanket searches and seizures of premises and persons without probable cause. Other human rights frameworks offer even more robust protections against police states. All these protections make policing less efficient, not more so.[8] Our usual standards of industrial accomplishment are an uneasy fit with a social practice like policing.

Software and AI are already informing much guard labor—that vast expanse of security (whether cyber or physical), policing, and military action that consumes a sizable proportion of global productive capacities.[9] There are plans to program robots that unerringly respect international law, unaffected by human bias, passion, or prejudice. Law professor Bennett Capers has imagined a future in which highly computerized policing will strike a blow for equality by subjecting the entire US population to extraordinarily high levels of surveillance.[10] In his Afrofuturist vision, law enforcement technology will be embedded in a gentler and more forgiving social order. Capers hopes that "what few trials are necessary will not quite be 'trial by machine,' but they may approximate it, since this too will eliminate the biases that tend to permeate trials in the present."

The Australian legal scholar and dean Dan Hunter believes AI could make prisons obsolete.[11] With a sophisticated enough computing system, all future sentences could be served as house arrest. Given the boredom and violence experienced by many in prison, that may seem like a great advance in humane punishment and rehabilitation. But Hunter's vision for a Technological Incarceration Project also includes proposals that could have been lifted from *Minority Report*. In exchange for serving out their sentences at home, "offenders would be fitted with an electronic bracelet or anklet capable of delivering an incapacitating shock if an algorithm detects that a new crime or violation is about to be committed. That assessment would be made by a combination of biometric factors, such as voice recognition and facial analysis."[12] This is the definition of automated justice. There is no chance to appeal to a judge or even to plea for mercy with an officer, once the AI adjudicates. It is judge, jury, and may well serve as executioner if the system miscalibrates its shock.

Technophiles offer particularly sharp and clarifying versions of the dominant policy approach to AI in law enforcement. In this vision, AI and robotics are inevitably taking over more of policing, and it is our duty—as citizens, scholars, and technologists—to help make it work as well as possible. For example: if facial recognition systems fail to recognize black faces as well as white, then add more black faces to the database. We should take a step back, though, and consider why universal surveillance or virtual-reality prisons seem so compelling to so many. Recall the lure of substitutive AI discussed at the beginning of this book: the worse or more expensively a job is now done by human beings, the more pressure there will be for machines to take it over.

When the violence and boredom of a poorly run prison is the only other alternative, constant monitoring at home by a taser-equipped robot doesn't look so bad. The shamefully racist record of so many police departments is what makes an Afrofuturist robo-cop seem like a great leap forward.

But what do we lose by buying into a logic of reformism when far more profound change is necessary? Princeton professor of African American studies Ruha Benjamin argues that "the purportedly more humane alternatives to prison, part of a growing suite of 'technocorrections,' should just be called what they are—*grilletes*" (Spanish for shackles)."[13] As police and prison wardens rapidly embrace AI and robotics, more holistic approaches to the problem of social control get lost in the shuffle. And the deeper harms inherent in any version of this technology go unaddressed.

What if the comparator for the *Minority Report* scenario was not a standard prison (cursed by cramped cells, uncomfortable beds, and bad food), but instead one of the "open prisons" common in Scandinavia? In these truly correctional facilities, governors (not wardens) provide inmates with sunny rooms and autonomy to cook their own food and clean their own living areas. There is education, non-exploitative work or job training, and a level of trust rarely if ever found in the prisons of the United States, China, and Russia.[14] These Scandinavian prisons are not country clubs—they still seriously restrict offenders' liberty. But they show a basic level of respect for criminals that is sorely lacking in both standard prisons and the technological "upgrades" now proposed for them. In policing, too, there is a growing movement to end, rather than mend, many troubling practices.

MISRECOGNITION AS DISRESPECT

At least two major British police departments have deployed face scanners to "to identify and monitor petty criminals, individuals with mental health issues and peaceful protesters." The NGO Big Brother Watch found that over 90 percent of the supposed "matches" were false positives.[15] That problem could in principle be solved by better technology or larger databases of, say, tagged photos from social networks.

However, there is yet another level of misrecognition going on here, which is not a technical problem. Why are the mentally ill lumped in with prisoners? Why does the state have a database of peaceful protesters? And finally, who authorized authorities to match already controversial classifications with technology designed to destroy remaining rights to obscurity—that is, to be anonymous in a crowd?[16] As machine vision spreads, critics are not only concerned about the efficacy of the technology. They are also questioning the power structures it supports and entrenches.

This misuse of facial recognition is growing in both the public and private sector. M.I.T researcher Joy Buolamwini has documented pervasive failures of commercial facial recognition (FR) software to identify minorities, and particularly minority women.[17] She calls this machine vision a "coded gaze," which reflects "the priorities, the preferences, and also sometimes the prejudices of those who have the power to shape technology."[18] Examples are not hard to find, even at the best-resourced firms. Amazon's Rekognition software made many false matches when the American Civil Liberties Union asked it to check whether members of the US Congress could be identified, by face, as criminals in its database. It also disproportionately identified minority members as criminals.[19] The Amazon system was so bad that it helped provoke several other large technology companies to call for regulation, despite the potential hit to their own bottom lines.

Firms are now working to ensure adequate representation of minority faces in relevant databases. But some civil rights advocates argue that the incompleteness of facial recognition databases is a good thing. Evoking a tragic American legacy of racist police brutality that continues to this day, academic and activist Zoé Samudzi argues that "it is not social progress to make black people equally visible to software that will inevitably be further weaponized against us."[20] Reducing some forms of bias in facial recognition systems (by, say, including more minority faces in them) may exacerbate others. Social Science Research Council president Alondra Nelson has said she's struggled "to understand why we want to make black communities more cognizant in facial recognition systems that are disproportionately used for surveillance."[21] Stopping false positives is an important goal. But the overall effects of facial recognition need to be addressed too, as legal scholar Woodrow Hartzog and philosopher Evan Selinger repeatedly warn in their interventions calling for bans on the technology.[22]

Futurist police consultants want to match most-wanted lists to universal closed-circuit TV cameras to nab roaming felons. Do we really want to enhance the ability of corporations and the police to match every face to a name—and an algorithmically generated, often secret, record?[23] It may be that facial recognition deserves the same reflexive "no" that greeted Google Glass. Google marketed the camera glasses as an easy way to augment reality, recognize acquaintances, and map unfamiliar spaces. But many of those caught in the gaze of Glass wearers found the camera glasses menacing. Awkward questions arose about Glass etiquette around urinals and bathrooms. As the technology failed to catch on, critics savaged its early adopters, calling them "Glassholes." Even a few decades ago, someone who brought a video camera everywhere would be dismissed as paranoid or obsessive. Should miniaturization (or near invisibility, in the case of a new generation of remote sensors) be enough to warrant a warmer response? No. In fact, unobtrusiveness is even more threatening, not even letting people know they are under watch. Moreover, biometric faceprints should be like fingerprints or bank account numbers—something that should not be fair game to just collect and disseminate.[24] Rather, a license should be needed, and regular audits by authorities imposed as a matter of course, to avoid misuse.

As biometric databases expand to cover ever more persons, the dynamics of power matter. In "Beijing's Big Brother Tech Needs African Faces," an essay for *Foreign Affairs*, Amy Hawkins noted that being "better able to train racial biases out of its facial recognition systems . . . could give China a vital edge," but also that "improving" this technology abets authoritarianism.[25] If South Africa had had the technological infrastructure that Beijing now deploys in the largely Muslim province of Xinjiang, would the anti-apartheid movement have developed?[26] It is hard to spin a narrative in which that apparatus helps a would-be Mandela.

Fortunately, there are well-developed methods of regulating dangerous (though sometimes useful) technology. Luke Stark of Microsoft Research has called facial recognition the "plutonium of AI." As Stark explains, "Plutonium has only highly specialized and tightly controlled uses, and poses such a high risk of toxicity if allowed to proliferate that it is controlled by international regimes, and not produced at all if possible."[27] For Stark, facial recognition also has so many negative consequences—from helping authoritarian regimes to laundering racial stereotypes with the mantle of science—that it deserves similar precautions and limitations. It may be reasonable for states to develop highly specialized and limited-use databases to stop pandemics or terrorism. But to deploy such powerful technology to ticket speeders, ferret out benefits fraud, or catch petty thieves is like using a sledgehammer to kill a fly.[28]

Stark's proposal is particularly insightful because it extends a widely agreed logic for limiting the power of machines: restricting violence. No one should be able to buy or equip an autonomous vehicle with machine guns on its hood without some kind of license; there's simply a common sense that tanks are for war, not personal armies. Many uses of facial recognition technology portend structural violence: systematic efforts to typecast individuals, to keep them monitored in their place, or to ransack databases for ways to manipulate them. Limiting of facial recognition technology ensures at least some freedom to move anonymously, with one's whereabouts and identity unmonitored by prying eyes.[29]

Unfortunately, in many societies, one of the horsemen of legal irresponsibility—free-expression fundamentalism—may upend sensible licensing schemes. Claiming a general right to "scientific freedom," firms have argued that no government should be able to tell them what they can or cannot collect data on or study. The great irony here is that the corporate assertion of constitutional rights creates databases that will have enormous chilling effects on ordinary citizens' speech. It is one thing to go to a protest when security personnel watch from afar. It is quite another when the police can immediately access your name, address, and job from a quick face scan purchased from an unaccountable private firm.

Privacy advocates have long worried about the accuracy of such databases. As Dana Priest and Bill Arkin revealed in their *Top Secret*

America series, there are hundreds of entities that enjoy access to troves of data on US citizens. Ever since the Snowden revelations, this extraordinary power to collate data points about individuals has caused unease among scholars, civil libertarians, and virtually any citizen with a sense of how badly wrong supposedly data-driven decision-making can go. There are high stakes; no-work, no-vote, no-fly, and no-citizenship lists all blackball individuals. Those unfairly targeted often have no real recourse and end up trapped in useless intra-agency appeals. The result is a Kafkaesque affront to basic principles of transparency and due process. Many regimes make it extraordinarily difficult for ordinary citizens to access—let alone challenge—the judgments governments and firms make about them. The moral case for "governance via AI" falls apart when such basic safeguards are missing.[30]

JUDGING A BOOK BY ITS COVER: FROM FACE
RECOGNITION TO FACIAL ANALYSIS

Facial analysis may soon make this already untenable situation much worse. Some prominent machine-learning researchers are claiming that our faces may reveal our sexual orientation and intelligence.[31] Using a database of prisoners' faces, others have developed stereotyped images of criminal features.[32] A start-up has claimed that it can spot the features of pedophiles and terrorists and that its technology is in use by security agencies around the world.[33]

Critics have challenged the accuracy of such predictions. Training data consisting of prisoners' faces are not representative of crime, but rather represent which criminals have been caught, jailed, and photographed. A high percentage of criminals are never caught or punished, so the data set cannot adequately represent crime itself. A study purporting to identify gay faces may have merely picked up on certain patterns of self-presentation of persons who use the dating sites that were the source of the "gay" and "non-gay" images used to train the classifier. Gay and lesbian persons *of a certain time and place* may be more or less likely to wear glasses, sport certain patterns of facial hair, or to present themselves smiling or more serious. As Dan McQuillan warns, machine learning often makes powerful predictions, "prompting comparisons with science. But rather than being universal and objective, it produces knowledge that is irrevocably entangled with specific computational mechanisms & the data used for training."[34]

All of these shortcomings support a larger critique of many opaque forms of machine judgment; being unexplained (or unexplainable), they stand or fall based on the representativeness of training data.[35] For example, imagine an overwhelmed court that uses natural-language processing to determine which of its present complaints are most like complaints that succeeded in the past and then prioritizes those complaints as it triages its workflow. To the extent that the past complaints reflect past conditions that no longer hold, they cannot be a good guide to which current claims are actually meritorious.[36] A more explainable system, which identified *why* it isolated certain words or phrases as indicating a particularly grave or valid claim, would be more useful. Even its value would decline as the types of cases filed, priorities of judges, or other factors changed.[37]

There are even more barriers to AI predictions of criminal activity. As Kiel Brennan-Marquez has explained, a jurisprudence of well-founded suspicion (largely arising out of Fourth Amendment law in the US) demands that authorities give a *plausible,* and not merely probabilistic, statistical, or artificially intelligent, account of the reasons why they investigate suspects.[38] We need to understand not simply *that* we are being watched, but also *why.* This is a limit on the power of the state, which may be all too tempted to use advanced surveillance technology to achieve complete control over citizens. Black box predictive analytics could easily give a police force an excuse to investigate nearly anyone, since we are all likely to have engaged in some behavior with *some* correlation with that of potential criminals.

This lack of causal connection (as opposed to mere correlation) points up another troubling aspect of face-based predictions. In the absence of evidence that face shape actually affects criminality, basing policy on mere correlation is creepy. It is effectively the elevation of an alien, non-human intelligence in a system in which human meaning and communication are crucial to legitimacy.[39] If we can actually tell people that there is some way in which changing their faces can reduce their likelihood of being criminal, then big data intervention presages exceptionally intense and granular social control. This double bind—between black box manipulation and intimate control—counsels against further work in the area.

The same caution should also govern applications of affective computing. The firm Affectiva prides itself on having some of the best "sentiment analysis" in the world. Using databases of millions of faces, coded for emotions, Affectiva says its AI can read sorrow, joy, disgust, and many other feelings from video of faces. The demand for such analysis is great. Beset by a backlog of manual security clearances, the US military is looking for AI that can flag suspect expressions.[40] There are numerous police, security, and military applications for emotion detectors, going all the way back to polygraph lie detectors (which are themselves the subject of significant controversy, and banned in US employment contexts).[41] Rana el Kaliouby, Affectiva's CEO and cofounder, has refused to license its technology to governments, but acknowledges that emotion recognition is a double-edged sword, wherever it is used. She imagines a workplace of the future with ambient monitoring of employees' emotional states:

> I do think it would be super-interesting if people wanted to opt-in, anonymously, and employers were able to then get a sentiment score, or just an overall view, of whether people are stressed in the office— or whether people are engaged and happy.
>
> Another great example would be where a CEO is giving a presentation, to people dialed in from around the world, and the machine indicates whether or not the message is resonating as they CEO intends. Are the goals exciting? Are people motivated? These are core questions that if we're all co-located, it would be easy to collect; but now, with everybody distributed, it's just really hard to get a sense of these things. However, if you turn it around and use the same technology to say, "OK. I'm going to pick on a certain member of staff because they seemed really disengaged," then that's a total abuse of the data.[42]

In el Kaliouby's vision, we can have the best of both worlds: pervasive monitoring without the data being used against us, and an AI understanding our feelings to serve us, without trying to master or manipulate us.

There is good reason to doubt both of those hopes. Employees in the US endure nearly "limitless surveillance," to quote the conclusion of one group of legal scholars.[43] Even in jurisdictions with robust privacy protections on the books, enforcement is often lacking. Nor have technology firms proven particularly good at policing the technology they license. Sure, ethical AI vendors may include language in their contracts that require employers not to pick on disengaged workers for failing to smile at the CEO's joke. But will they be able to crack into trade secret–protected methods of employee evaluation? What is the business case for a firm to continually police uses of its products? Even in jurisdictions where such invasive surveillance would violate labor laws, enforcers are scarce and preoccupied by more immediate threats. Third-party auditors are no panacea either. They have failed repeatedly even in settings where law enforcement demanded strict scrutiny of corporate practices.

In all the cases mentioned above, we need to do more than make vague commitments to corporate social responsibility and regulation. We need to structure communities of practice where workers are empowered to whistle-blow and to push back against abusive uses of technology (whether within or outside their firms). We need independent and well-funded regulatory bodies capable of enforcing clear rules. And we need real penalties for breaking them. These are the essential structural foundations for humane automation. The debate over the terms and goals of accountability must not stop at questions of fairness. We must consider broader questions, such as whether these tools should be developed and deployed at all.[44]

WHEN FINANCIAL INCLUSION BECOMES CREEPY AND PREDATORY

The divide in debates over facial recognition and analysis—between those who want to increase AI's predictive capabilities and those who want to ban or severely limit its use—will arise repeatedly in finance as well as law. Lenders rely on data to grant credit and set interest rates. They constantly demand more information from would-be borrowers, to sharpen the predictive powers of their algorithms. The data demands can become so far-reaching, intrusive, and speculative that they begin to feel dehumanizing. Where to draw the line?

The promise of machine learning is to find characteristics of good risks, as identified by past lending, that might be shared by people now locked out of lending. For example, it may turn out that good credit risks tend to get at least seven hours of sleep a night (as roughly measured by phone inactivity), use Google Chrome as their web browser, and buy organic food. Individually, none of these correlations may be strong. Once they reach a certain critical mass, a new and dynamic big data–driven profile of the good credit risk may emerge. Micro-lenders are already using traits like internet activity to decide on loan applications.

Lenders say all these innovations are a good reason to further deregulate finance. Why worry so much about bias when AI is just a computer program, unbeholden to human emotions and prejudices? And why try to look out for consumers when AI gets better and better at determining who can repay a loan? Emboldened by fintech utopianism, regulators around the world are loosening the reins for new firms. The United States and Hong Kong have endorsed "regulatory sandboxes" for fintech firms, limiting scrutiny. This is a misguided approach, since the use of AI in fintech has just as many problems as traditional underwriting, if not more. It leads to a "scored society" where individuals lack basic information about how they are being judged.[45]

These problems are troubling in the abstract. Their concrete implications are chilling, as one Privacy International report revealed. Fintech firms have already scored creditworthiness based on political activity, punctuation, and assumed sleeping patterns. According to the NGO Privacy International, "If lenders see political activity on someone's Twitter account in India, they'll consider repayment more difficult and not lend to that individual."[46] Always on the lookout for potential correlations, predictive analytics finds meaning in even the most mundane actions: "[One firm] analyses the way you fill in a form (in addition to what you say in the form), and how you use a website, on what kind of device, and in what location."[47] In China, sharing "nice messages about the government or how well the country's economy is doing, will make your score go up."[48] A US fintech firm says that for it, filling in one's name on forms with capital letters is a warning sign.[49]

We all may be emitting other "tells" that are more grave. For instance, researchers recently parsed the mouse movements of people who extensively searched for information about Parkinson's disease on Bing.[50] Some in this group—which is far more likely to have Parkinson's than the population as a whole—tended to have certain tremors in their mouse movements. Such tremor data, and similar physical activities like speed of typing and mouse movements, are not the kind of performance we expect to be judged on. As more databases are combined and analyzed, other, even more subtle signals about our likely future health conditions will emerge. The more data about precursors to troubling fates is available, the better AI will get at predicting them.

We may want our doctors to access such information, but we need not let banks, employers, or others use it. There is more to life—and public policy—than consenting transactions among borrowers and fintechs to ensure "financial inclusion" (the industry's go-to rationale for deregulation). Without proper guardrails, there will be a race to the bottom in both data sharing and behavior shaping as more individuals compete for better deals. That would result in a boom in predatory inclusion (which harms more than it helps), creepy inclusion (which gives corporate entities a voyeuristically intimate look at our lives), and subordinating inclusion (which entrenches inequality by forcing persons to maintain the same patterns of life that resulted in their desperation in the first place). Lawmakers should discourage or ban each of these types of "inclusion."

Predatory inclusion is a concept with a long history.[51] Credit enables, while its shadow side (debt) constrains. Sometimes the burdens of loans are far greater than their benefits. Imagine a couple, Joe and Margaret, who take out a payday loan with a 20 percent per week interest rate. At that rate, the debt more than doubles every month. Desperate to repay it, they have no money for dinner for their children for a few days. They fall behind on rent and get an eviction notice, but the lender is paid back in full within three months and makes a tidy profit. When we think about the future of AI in consumer finance, much depends on how an experience like Joe and Margaret's is coded by the firms now making massive profits from millions of personal liquidity crises. Crude metrics of repayment will incentivize machines to keep seeking out such desperate borrowers.

The elements that make for a "successful" transaction matter. Lenders focused on the bottom line rarely, if ever, record data on whether repayment has been easy or arduous, humiliating or routine. Ignorance is bliss—for the lender. Borrowers suffer in silence. One Kansas war veteran paid $50,000 in interest on $2,500 in loans.[52] A British nursery worker borrowed £300 and found herself owing £1,500 a year later.[53] Hundreds of borrowers told their own harrowing stories to London's the *Guardian,* and American tales of woe are even more common. In the developing world, the stakes can be even higher. Stories of suicides among clients of micro-lenders unable to repay their loans are not uncommon. "Credit circles" mean that fellow villagers of the debtor will often be saddled with his debt, creating unbearable shaming dynamics. One Kenyan argued that digital lending apps are "enslaving" both the working poor and the salaried classes in his country.[54]

Predatory inclusion also poisons education's promise. When those desperate for opportunity take on a heavy loan burden to attend training programs of dubious value, the latter effect predominates. The rhetoric of uplift convinces too many that more learning is a sure path to better earning power. Fly-by-night for-profit colleges take advantage of their hope.[55] They peddle a cruel optimism—that the future has to be better than the past.[56]

One way of preventing exploitative educational debt is to certify which programs have provided a good "return on investment" to students. In the United States, the Obama administration agonized for years over such a requirement, finally promulgating it as a "gainful employment rule" in 2015—only to see the next administration immediately roll back those protections. There was little public support for the Obama-era rule, a hyper-technocratic approach toward certifying programs for loan eligibility. Its obsessive focus on expected earnings soon after graduation implied that the entire point of education was reaching a certain earning potential. Whatever the merits of such an approach, it could not be applied to debt in society as a whole. No government can comprehensively specify what credit can be used to purchase, and what is forbidden.

Other jurisdictions have taken a more enlightened approach to educational debt, which could be applied in many other credit contexts.

For example, income-based repayment programs specify a percentage of income due for just a certain number of years.[57] It is essentially a graduate tax—except that those who make enough income to pay off the loan balance stop paying once they do so.[58] The "free college" movement has also demanded direct provision of education.[59]

By this point, we may seem to have come a long way from AI in finance. But the tech ethics community is becoming increasingly aware that AI's chief problems are not primarily technological. Rather, the social role of AI is critical. If we want to stop predatory lending, addressing the power imbalance between creditors and debtors is more important than tweaking AI technology.

The same principle also applies to "creepy inclusion." Nonstop cell phone tracking, archiving, and data resale is a prime example of "creepy inclusion."[60] If a banker asked an employee if she would mind his trailing her twenty-four hours a day, seven days a week, the feeling of threat would be imminent. She might even be able to get a restraining order against him as a stalker. Although credit contracts that come with cell phone trackers may seem like less of a threat, the use and reuse of data creates a distinctive and still troubling menace. Creepiness is an intuition of future threat, based on a deviation from the normal. Normal experience includes entitlement to a work life distinct from home life and having judgments about us being made on the basis of articulable criteria. Creepy inclusion disturbs that balance, letting unknown mechanical decision makers sneak into our cars, bedrooms, and bathrooms.

At present, such surveillance demands are rare. Financial sector entrepreneurs brush aside calls for regulation, reassuring authorities that their software does not record or evaluate sensitive data like location, person called, or the contents of conversations.[61] Metadata, however, is endless, and as we saw with the example of the hand tremors predicting Parkinson's, can yield unexpected insights about a person.[62] Now is the time to legislate against creepy inclusion, before manipulative marketing tricks so many people into bad bargains that industry can facilely assert its exploitation is a well-established "consumer preference."

The timing issue is critical, because industry tries to deflect regulation when a practice first begins, by deeming it "innovation." "Wait and see how it turns out," lobbyists say. Once a practice has been around for a while, another rationale for non-regulation emerges: "How dare you interfere with consumer choice!" This cynical pair of rationales for laissez-faire is particularly dangerous in data regulation, since norms can change quickly as people jostle for advantage.[63] Uncoordinated, we rapidly reach an equilibrium that forces everyone to reveal more to avoid disadvantages. Cooperating to put together some enforceable rules, we can protect ourselves from a boundless surveillance capitalism.[64] For example, some jurisdictions are beginning to pass laws against firms micro-chipping workers by subcutaneously injecting a rice-sized sensor underneath their skin.[65]

That project of self-protection is urgent, because subordinating inclusion is bound to become more popular over time. Penalizing people for becoming politically involved—as some firms in India and China now do—further entrenches the dominance of those who provide credit over those in need of it. The rise of what Citibank has called "vox populi risk"—including the supposed "danger" of persons demanding that corporations treat them better—will provoke more executives to consider the political dimensions of their lending. Firms may decide that those who sue their landlords for breach of a lease, or file a grievance at work, are more costly customers, since they are more likely than average to assert their rights. Such criteria cannot inform a humane credit system. They set us all on a contest of self-abasement, each of us eager to prove him or herself the type of person who will accept any indignity in order to get ahead.

While predatory, creepy, and subordinating inclusion are objectionable on diverse grounds, they all clarify a key problem of automation. They allow people to compete for advantage in financial markets in ways that undermine their financial health, dignity, and political power. As the third new law of robotics suggests, it is critical to stop this arms race of surveillance now, before it is normalized.

INTERNALIZING OBEDIENCE IN A SCORED SOCIETY

In 2014, China's State Council announced an "Outline for the Promotion of a System of Social Credit," which includes scoring as one component of reputation assessment.[66] Social credit may affect all manner of opportunities, ranging from the right to travel, access to loans and schools (for oneself and one's children), and much more. The State Council has described an overarching plan "to encourage sincerity and punish insincerity," which in 2014 it asked "all provincial, autonomous region and municipal People's Governments, all State Council ministries and commissions, [and] all directly subordinate departments" to "implement earnestly."[67]

Many local programs have developed since then. In Yangqiao, residents "can earn a maximum rating of 3 stars in a 'Morality Bank' program . . . based on attributes such as filial piety or good deeds."[68] The rating can mean better terms on loans. In Rongcheng, citizens have a "base score of 1,000 per citizen or company, along with a letter grade ranging from AAA to D," which can be affected by various reports of untrustworthy, insincere, or deviant behavior, whether explicitly proscribed by law or merely considered "uncivilized" by authorities.[69] In Ningbo, "dodging fares on public transport or delaying payment of electricity bills" could "affect an individual's chances of getting a mortgage, a promotion or a raise."[70] Larger cities, like Beijing, will soon be ready to deploy their own scoring projects. In its most ambitious forms, social credit scoring promises the automation of virtue—or at least its measurement. Citizens will receive real-time feedback on their relative rank, free to compete on terms set by bureaucrats.

Social credit systems (SCSs) in China may be based on myriad data points. None appear to have yet integrated the full scope of the state "outline." However, authorities are clearly advancing evaluative systems with three particularly troubling characteristics. First, SCSs (whether nationally or locally implemented) are poised to integrate a *comprehensive* or near-comprehensive range of data—from postings online to traffic violations, from home life to work conduct, and much more. Second, blacklists (censuring individuals) and redlists (commending them) mean that there is *rippling* impact well beyond the original source of violation.[71] That is, a failure to pay a debt may not only affect one's financial credit and status, but may also ripple beyond that to affect one's ability to travel, receive government benefits, or enjoy other privileges of citizenship. Third, disrepute can have *networked consequences*, affecting the scores of family or friends of a "discredited" person.[72]

These three elements of SCSs (comprehensiveness, rippling, and net-worked consequences) have the potential to be a perfected means of social control, particularly if they apply to communication about rights, politics, and SCSs themselves.[73] Rippling means that deviance may ramify into years of repercussions. Advancing biometrics (in iris, facial, and gait recognition, among others) make the adoption of a new persona almost impossible to accomplish once stigma has spoiled one's identity. Even those who are brave enough to face the consequences of disobedience themselves would likely be loath to risk imposing burdens on those they care about.[74] The ripple effect of blacklisting or score lowering makes it difficult, if not impossible, to assess the full impact of a violation of law. Redlisting may be even more granularly oppressive, inculcating in citizens an "algorithmic selfhood" attuned to maximizing status.[75]

The Chinese government has claimed that SCSs simply reflect the values now embodied in Chinese families, schools, and courts. But with no standard appeal mechanism, automated distribution of stigmatizing blacklists (as well as commendatory redlists) threatens to supplant, rather than supplement, the authority of families, schools, and courts.[76] Aspects of the SCS could easily end up serving as a quant-driven power grab, enabling bureaucrats to assert authority over vast swathes of social life in a way they could never achieve via public legislation. Such quantitative governance of culture is a paradox. The very effort to articulate the precise value of manners, emotions, and online messages undermines their authenticity, as spontaneous affections and interactions are instrumentalized into point scoring. This is one of many perils of formalizing evaluative criteria in ineffable realms of family, friendship, community, and manners.

Legal scholars are already raising concerns about SCSs' accuracy and fairness. There are already worthy efforts afoot to ensure due process for the rewards and punishments they mete out. It is crucial to note, however, that if reformers focus only on legalistic concerns, their push for algorithmic accountability may miss the forest for the trees. Social credit systems are as much cultural and political projects as they are legal ones. Algorithmic governance allows government and corporations (often working in tandem) to consolidate and thus dominate spheres of reputation that should be more decentralized and private.[77] It is prone to a tyrannical granularity, exacting and intrusive.[78]

Sadly, the rush to monitor and measure goes well beyond SCSs. American firms are using opaque and unaccountable AI for behavior assessment.[79] A global edtech industry has pushed for behavioristic testing and ranking of students, schools, and teachers.[80] The same monitoring technologies may also dominate hospitals, nursing homes, and daycare facilities. Wherever the Taylorist impulse to measure and manage prevails, such methods may spread. Reputational currencies rebrand repression as rational nudging.

TURNING THE TABLES ON JUDGMENTAL AI

There are a dizzying array of AI systems that claim to score, rank, or rate human beings. Those judged wanting are now turning the tables and holding AI itself up to ethical standards. From SCSs and fintech to facial analysis and predictive policing, judgmental AI demands too much data. It makes opaque, unjustifiable, or unfair inferences. While promising security and inclusion, it all too often delivers precarity and stigma. And these are merely the individual-level harms. Just as important in the long run are the ways that AI overreach threatens to erode the norms and values that hold together institutions.

Following philosopher Michael Walzer, we might categorize the family and many civil-society institutions as distinctive "spheres of justice"—realms of human experience that ideally operate according to their own logic of distribution (of both recognition and resources), rather than standards externally imposed (whether by bureaucrats or AI).[81] Walzer advanced the spheres of justice ideal as a way to rescue and revive philosophical approaches not encompassed by utilitarian and deontological theories dominant in Western ethical theory.[82] He engages in a thick description of the normative foundations of current social practices in varied spheres, with openness to good arguments for change or reform. There will be debates over what exactly makes a person a good child or parent, religious believer or cleric, gardener or conservationist, boss or worker. But one core idea of Walzer's is that one's standing in one sphere should not be unduly affected by one's conduct in another. In other words, simply because a person goes bankrupt or speeds, that should not (itself) seriously reduce his reputation as an employee, minister, or parent, much less that of his relatives.

These ideas may sound abstract. But taken seriously, they would upend many of the big data methods explored in this chapter. Such a principle condemns both the "rippling" effect of social credit scoring and the "big data hiring" that uses facial expressions, gait, and internet search history to evaluate a job applicant. Why should a person be charged a higher rate of interest merely because she bought a brand of beer frequently purchased by defaulters? In the realm of black box AI, this makes perfect sense—what matters is prediction and correlation, not justice or fairness. If, however, we consider credit a sphere of justice—something implicit in all the moral discourse surrounding obligation, bankruptcy, and promises—the correlationist impulse makes little sense. A person deserves to be judged on her own merits, in a way that is publicly intelligible and legitimate.[83]

Why do these spheres of justice deserve respect, compared with government or market actors who may be vastly more technologically sophisticated? Jürgen Habermas's concept of the "systemic colonization of the lifeworld" buttresses Walzer's case for a separation of spheres.[84] For Habermas, the bureaucracies of both governmental and market actors are always in danger of over expanding, "juridifying the lifeworld" by imposing rules for correct conduct that oversimplify, distort, or outright overrule extant ideals.[85] These systems of commerce and governance organize our lives, for the most part, behind our backs, via market exchanges or politico-bureaucratic decisions. The comparative legitimacy of the lifeworld of civil society, the family, grassroots institutions, and other more human-scale interactions, inheres in the feeling of some degree of phenomenological immediacy, apprehensibility, and control surrounding our conduct within them.

At its worst, systemic colonization of the lifeworld encourages a kind of world- and self-instrumentalization that downplays (or entirely eclipses) intrinsic values.[86] "You cannot manage what you cannot measure" is a mantra of bland neoliberal managerialism, an assumption that observations that are not comparable and computable are not as valuable as those that are. Like the instrumentalizing "male gaze" critiqued by feminists, a "data gaze" now threatens misrecognition and reification.[87] Unmoored from some concrete sense of meaning and purpose, the data gaze undermines not only our own autonomy, but also the integrity of the institutions that give that autonomy meaning.

There must be limits on corporate and governmental shaping of human behavior, minds, and souls. Ubiquitous sensor networks and AI have given us a world in which every act could count for or against each consumer and citizen in a "reputational bank." In the United States, the "balances" in these reputational accounts are fragmentary and dispersed into thousands of disconnected data centers, often in secret. Chinese social credit scores promise to be more centralized and their criteria of evaluation more public. Neither approach is satisfactory. Secret scores are fundamentally unfair; they give the scored person no chance to contest them. When detailed scoring systems are public, they threaten to regiment behavior in excessive and intimate ways. We must therefore go beyond efforts to merely reform scoring, and instead turn to limits on its scope, judging AI itself as falling short in crucial ways.

AI is not the only way of ordering society and judging who should receive benefits and burdens. There are numerous other practices of expertise, evaluation, care, and concern. There are distinctive patterns of human conduct that can be performed better or worse, more or less skillfully, but whose quality is better described and prescribed in a narrative, holistic, and emotive mode rather than reduced to numbers or coded into instructions. Thus, one challenge for critics of "machines judging humans" is the maintenance, improvement, and further development of non-quantitative valuation—that is, judgment. A robust critique of AI evaluation depends on some positive account of the older methods of judgment that AI methods are now eclipsing.[88] For example, in the academy, tenure letters and Festschrift tributes are an alternative form of evaluation to citation counts and impact factors. Essays explaining the shape of one's career and one's reasons for choosing a certain topic or method of research are a type of self-evaluation that should become more popular among scholars. We need better, more narrative, mid-career assessments of the depth and breadth of scholarly contributions. Such qualitative modes of evaluation can be far richer than the quantification-driven metrics now ascendant in the academy.

In short, it "takes a theory to beat a theory," and an alternative method of explaining what professionals do and how one can do it in better or worse ways is necessary to fend off the temptations of AI evaluation. Such narratives may impose their own disciplines and anxieties. But at least they relieve us from the fantasy that scholars, doctors, nurses, teachers, and professionals of all kinds can be judged along some commensurating metrics calculated via machine learning. They could also serve as examples of humane judgment in many other contexts.

PRODUCTIVITY VS. POWER

In a far-reaching intellectual history, Will Davies describes a broad shift in the paradigm use of knowledge in society.[89] Enlightenment intellectuals thought knowledge should create a shared picture of reality—both scientifically (more accurate models of the world) and socially (to create a public sphere based on a shared understanding of certain key facts, values, and norms). However, the rise of economic thought portended another vision of knowledge, whose key value lay in attaining an advantage over others. Perhaps necessary in commerce, it has had an insidious impact in the public sphere, education, finance, and beyond, as the acquisition of knowledge to gain power crowds out other, more public-regarding purposes.

As firms and governments increasingly turn to machines to judge persons, they grant enormous power to those developing AI. Power is the ability of one person to make another do something the latter would not do otherwise.[90] It is most obvious in politics and war, but power matters in the economy, society, and families as well. Artificial intelligence may both entrench and disrupt existing power relations in schools, workplaces, and even families. We need to be cognizant of this dynamic and to guard against its most destructive manifestation: arms races.

We can all intuitively understand how two military rivals waste resources as they stockpile missiles, missile defense systems, ways to jam or evade missile defense systems, and so on. It is individually rational to keep ahead of one's enemies, but it is collectively insane for all to endlessly try to outdistance every other. There is no objective amount of money one can spend in order to be "safe."[91] There is only relative advantage, which constantly risks being upset if an enemy comes up with a new tactic or technology. That is one reason the United States, despite paying more money for "defense" than the next seven countries combined, keeps investing more money in military, police, and intelligence services.

Arms race models are relevant well outside of military contexts. In *The Two-Income Trap,* then-professor Elizabeth Warren described middle class families bidding up real-estate prices in better school districts. Economist Robert Frank's *Falling Behind* offered a general theory of such arms races. When there is a limited supply of something—such as power, prestige, or prime land downtown—competition is inevitable. In many of these arms races, law and finance powerfully influence the ultimate victor; the person who can attain the largest mortgage, for example, can outbid others on a house.[92]

Law, politics, and even political advocacy can also degenerate into arms races. Frank describes commercial litigation as a money pit, where each side pours resources into law firms (and now legal analytics) to wrest an edge over its opponent. In politics, even the smallest edge is critical; a candidate wins an election not by getting some magic number of votes, but simply by getting *more* votes than opponents. Such edges are not simply set back to zero once a new election rolls around. The majority party (sometimes by a single vote) can work its will to promote its allies' interests and disadvantage its enemies. In some countries, democracy itself erodes after a few rounds of self-reinforcing advantages accrue to winners. Political campaigns can feel like war by other means: the battle for mindshare is zero-sum.[93]

The third new law of robotics—discouraging AI-driven arms races—should inform all these fields. It is a common thread of this chapter and the next. In the name of rationalizing social judgments of hiring, firing, criminality, and more, scoring AI threatens to put us all on a slippery slope toward competing for status by revealing all aspects of our lives to powerful and opaque entities, to be regimented according to their ends. Only concerted action can stop this race toward a full disclosure future.

Sometimes the path toward cooperation will be relatively clear—as, for instance, in financial regulation that would limit the ability of lenders to use drones or social media analysis to observe the behavior of current or would-be customers. In other situations, particularly great-power military competition, our ability to impose restraint may rely on little more than fragile norms or international relations. Nevertheless, humane strategies of cooperation are essential, lest ever more details of social order are relegated to machines.

6

▪ Autonomous Forces

In the classic Atari game *Pong*, a player moves a "paddle" (signified by little more than a thin rectangular block on the side of the screen) in order to defend against an incoming ball and to hit the ball past the opposing player's paddle. One of the simplest video games ever developed, *Pong* reduces table tennis to two dimensions. To win requires dexterous hands and a good intuitive understanding of geometry.

Or at least, that is what it takes for humans to win at *Pong*. A team of AI researchers tried a radically different approach. They simply set a computer to try every imaginable response to the incoming ball—avoiding it, knocking it directly back, hitting it with the edge of the paddle, all at slightly different angles or speeds. For an AI capable of trying millions of strategies a second, optimal playing patterns quickly emerged. The AI mastered *Pong* and was able to defeat any human player. It later learned how to beat anyone at other video games, and even the ancient Chinese board game of Go.[1]

AI researchers called these wins a major breakthrough, because they resulted from self-learning. The program did not study past games of *Pong* or Go to glean strategies, as a human might do. Rather, by a combination of brute force (simulating a massive array of scenarios) and algorithmic ordering (of the optimal response to each), the AI was able to master the game. There does not appear to be a way for a human to beat it.

That kind of domination is a fantasy of military theorists, who have long modeled war games to simulate enemy moves and countermoves. Indeed, the field of AI has roots in the cybernetics of the mid-twentieth century, when operations research experts advised generals on how best to program automated responses to an enemy presumed to be hellbent on its own technological advance.[2] The modeling has a hall of mirrors quality—of warriors trying to project what the enemy projects they will be projecting, the better to surprise them.

We've already discussed the unique ethical and legal difficulties presented by the robotization of policing. The use of AI in war adds yet another layer of complexity. All the arms-race scenarios I have described so far have presumed the existence of a state that can set rules and punish those who break them. Globally, no such power exists. The United Nations can condemn a state, but its authority is often flouted.

The lack of global governing authority turns the discussion of robots in war into a game of paired antinomies. Abolitionists seek to outlaw killer robots via international treaties. Self-described "realists" argue that states must stockpile advanced military technology, lest they be overrun by less scrupulous competitors or terrorized by brigands. AI boosters claim that robots will reduce the horror of war, targeting force better than any person could. Skeptics think that that future is a long way off. Deterrence theorists worry that even if automated war is more "humane," armed conflict must not be made too easy for strong states, lest they use their technical advantage to dominate everyone else.

After exploring the virtues and limits of abolitionist and realist approaches to military AI, this chapter recommends forms of attribution that, pursuant to the fourth new law of robotics (requiring identification of robot controller and owner), may make it easier for states to abide by the second law (discouraging arms races). Arms control may be a fraught and dangerous process, but it will be less so if nations can work together to at least provide an honest accounting of their capacities and actions. We can also learn from the history of nuclear nonproliferation. There are far fewer nuclear-armed states than there are governments with the capacity to make such weapons. The story of nuclear paths not taken illuminates a way out of the labyrinth of AI strategy now bogging down concerted global action against an AI arms race.

IMAGINING ROBOTIZED SLAUGHTER

The video is stark. Two menacing men stand next to a white van in a field, holding remote controls. They open the van's back doors, and the whining sound of quadcoptor drones crescendos. They flip a switch, and the drones swarm out like bats careening out of a cave. In a few seconds, we cut to a college classroom. The killer robots flood in through windows and vents. The students scream in terror, trapped inside. The film *Slaughterbots* leaves the rest to our imagination, but the import is obvious. Palm-sized killer robots are either here, or a small technological advance away. Terrorists could easily deploy them. And existing defenses are either weak or nonexistent.[3]

After the Future of Life Institute released *Slaughterbots*, some leading lights in the defense community complained. They argued that the film sensationalized a serious problem, stoking fear where sage reflection was required. But the line between science fiction and industrial fact often blurs in war futurism. The United States Air Force has predicted that "SWAT teams will send mechanical insects equipped with video cameras to creep inside a building during a hostage standoff."[4] One "micro-systems collaborative" has already released Octoroach, an "extremely small robot with a camera and radio transmitter that can cover up to 100 meters on the ground."[5] If a waterbug can enter a crack in the wall, so too could the Octoroach. Who knows how many other noxious creatures are now models for drone "swarming" technology, another watchword of avant-garde military theorists.

Peter Singer and August Cole's novel of technologized war, *Ghost Fleet*, presented a kaleidoscopic vision of autonomous drones, hijacked satellites, and lasers deployed in a war pitting the United States against Russia and China.[6] The book cannot be written off as a techno-military fantasy: it includes hundreds of footnotes documenting the development of each piece of hardware and software it describes. Both authors have presented at military venues.[7]

Advances in the theoretical modeling of robotic killing machines may be even more disturbing than trends in weaponry. A Russian science fiction story from the 1960s, "Crabs on the Island," described an iterated *Hunger Games* for algorithmic agents, where robots would battle one another for resources. Losers would be scrapped, and winners would spawn, until some evolved to be the best killing machines.[8] When a leading computer scientist mentioned a similar scenario to the US's Defense Advanced Research Projects Agency, calling it a "robot Jurassic Park," contacts there called it "plausible."[9] It does not take much thought to realize that such an experiment has the potential to go wildly out of control.[10] Part of becoming a perfected killing machine is knowing how to evade capture by others, be they humans or robots. Expense is the chief impediment to a great power experimenting with such a project. Software modeling may eliminate even that barrier, allowing virtual-battle-tested simulations to inspire future military investments.

There is ample precedent for states to unite around prohibitions of particularly gruesome or terrifying weapons. By the mid-twentieth century, international conventions banned both biological and chemical weapons. The community of nations has banned the use of blinding laser technology as well. A robust network of NGOs has successfully urged the United Nations to convene member states to agree to a similar ban on lethal autonomous weapons technology (LAWS). And while there have been (and will be) lengthy arguments about the definition of LAWS, we can all imagine some subset of particularly terrifying *kinds* of such weapons that all states should agree never to make or deploy. For example, a drone that gradually heated enemy soldiers to death would violate international conventions against torture.[11] Similarly, sonic weapons designed to wreck an enemy's hearing or balance should merit similar treatment. A designer of such weapons, like a designer of a deadly flu virus or noxious gas, should be exiled from the community of nations.

THE UNEASY CASE FOR KILLER ROBOTS

Is a killer robot as horrific as biological weapons? Some military theorists say they are not merely technologically superior to old weapons, but also more humane. According to Michael Schmitt of the US Naval War College, autonomous weapon systems (AWS) offer new and better ways to target attacks to minimize casualties. For instance, given likely advances in face- or gait- recognition technology, a killer robot might only shoot men between twenty-one and sixty-five in a village.[12] Schmitt also envisions autonomous weapons systems as guarantors of peace. He argues that they could "police the skies" to ensure that a slaughter like Saddam Hussein's killing of Kurds and Marsh Arabs could not happen again.[13]

For establishment military theorists and computer scientists, programmed weapons offer the best promise of balancing the demands of war against the norms of international humanitarian law. Ronald Arkin of the Georgia Institute of Technology believes that AWS may "reduce man's inhumanity to man through technology," since a robot will not be subject to all-too-human fits of anger, sadism, or cruelty. He has proposed taking humans out of the loop of targeting decisions, while coding ethical constraints on lethal actions into robots. Arkin has also developed coding of targets (to, say, exempt hospitals).[14]

Theoretically, Arkin has a point. It is not easy to imagine a robotic My Lai massacre. Such cruelty seems rooted in forms of human irrationality that are more emotional than calculative. Yet we often reserve our deepest condemnation not for violence done in the heat of passion, but for the premeditated murderer who coolly planned his attack. It is also hard to imagine a robotic weapons system without some kind of override feature for various of its limitations, which would of course be controlled by some human soldier.

Any attempt to code law and ethics into killer robots raises enormous practical difficulties. Computer science professor Noel Sharkey has argued that it is in principle impossible to program a robot warrior with reactions to the infinite array of situations that could arise in the heat of conflict.[15] The curse of dimensionality is too strong, particularly in an arena where, blessedly, we do not have an enormous amount of past data to guide future action. Like an autonomous car rendered helpless by snow which interferes with its sensors, an AWS in the fog of war is dangerous. Machine learning has worked best where there is a massive dataset with clearly understood examples of good and bad, right and wrong, decisions. For example, credit card companies have improved fraud detection mechanisms with constant analyses of hundreds of millions of transactions, where false negatives and false positives are easily labeled with nearly 100 percent accuracy. Would it be possible to "datafy" the experiences of soldiers in Iraq, deciding whether to fire at ambiguous enemies? Even if it were, how relevant would such a dataset be for occupations of, say, Sudan or Yemen (two of the many nations with some kind of US military presence)?[16]

The premise of the big data driving ordinary successes in predictive analytics is a high volume of readily coded information. Most counterinsurgency soldiers would testify that the everyday experience of war is long stretches of boredom punctuated by sudden, terrifying spells of disorder. Standardizing accounts of such incidents is a challenge. Narratives of encounters in military occupations are not always available—and seem critical to parsing whether any given situation is properly characterized as handled well or poorly. Even more troublingly, the "weaponization of information" is an increasingly important strategy of conflict—as is the deployment of misinformation.[17] From Vladimir Putin's revival of the classic Russian strategy of *maskirovka* (military deceptions), to the US Army's reports on strategic uses of misinformation, militaries globally realize that the framing of conflict is just as important as military action itself.[18] Given the paucity of data about conflict and its susceptibility to manipulation, aspirations for ethical robots appear unrealistic.

KILLER ROBOTS AND THE LAW OF WAR

International humanitarian law (IHL), a set of norms governing armed conflict, poses even more challenges to developers of autonomous weapons.[19] A key norm of IHL is the "rule of distinction," which requires warring parties to distinguish between combatants and noncombatants. Only combatants can be targeted.[20] There is no solid convention of "self-identification" as combatants (like uniforms or insignia) among guerillas or insurgents. Instead, in counterinsurgency and other unconventional warfare that has become increasingly common over the past few decades, combatants blend in with civilians. The NGO Human Rights Watch offers the following example of a potential machine-driven tragedy, preventable by human judgment:

> [A] frightened mother may run after her two children and yell at them to stop playing with toy guns near a soldier. A human soldier could identify with the mother's fear and the children's game and thus recognize their intentions as harmless, while a fully autonomous weapon might see only a person running toward it and two

armed individuals. The former would hold fire, and the latter might launch an attack.[21]

Given the present limits to programming and facial recognition, this is a powerful criticism of robotics in war. A key ethical principle of warfare has been one of discrimination: requiring attackers to distinguish between combatants and civilians.

Proponents of AWS, however, insist that such weapons' powers of discrimination are only improving. Once perfected, drone surveillance might enable "combat zones that see," meticulously keeping track of who among the enemy are armed and dangerous and who have always been quiescent, offering no suspect signals or unusual patterns of activity.[22] Even if we presume the technology will become more targeted, there is still a massive leap in logic to assuming that commanders will buy it or use it or develop just principles of discrimination in the fog of war. The category of "combatant" (a legitimate target) has already tended to "be diluted in such a way as to extend to any form of membership of, collaboration with, or presumed sympathy for some militant organization."[23]

The principle of distinction is only one of many international laws governing warfare. The Geneva Conventions' "rule of proportionality" prohibits "an attack which may be expected to cause incidental loss of civilian life, injury to civilians, damage to civilian objects, or a combination thereof, which would be excessive in relation to the concrete and direct military advantage anticipated."[24] Adjudicating "combatant" status may, in principle, be within the reach of a panoptic precursor to Skynet, which can track all persons in a territory to assess whether they are armed or have otherwise participated in hostilities. But even the US Air Force has called the determination of proportionality "an inherently subjective determination that will be resolved on a case-by-case basis."[25]

Several authorities have explained how difficult it is to program a robot to handle the "infinite number of scenarios it might face" in war.[26] Peter Asaro, an expert on robotics trained in both computer science and the philosophy of science, argues that a "utilitarian calculus of risk minimization" cannot substitute for the kind of legal analysis characteristic of lawyers' past efforts to interpret the Martens clause.[27] Both military officials and international tribunals should instead address the wide array of ethical concerns raised by increasingly autonomous weaponry. Asaro identifies the stark incongruity between robotic data processing and the human reason characteristic of juridical processes. As he observes, "An automated process designed in advance, based on sensor data, is neither a legal judgment nor a moral judgment."[28] However effective targeting technology is at monitoring, detecting, and neutralizing threats, there is no evidence that it can engage in the type of subtle and flexible reasoning essential to the application of even slightly ambiguous laws or norms.

Historian Samuel Moyn adds another layer of moral concern here. Even if we were to assume that technological advances in war robotics were targeting force to make war less lethal than ever, would that be a good thing? Surveying the advancing influence of human rights principles on the law of conflict, Moyn observes a paradox: warfare has become both "more humane and harder to end." For invaders, robots spare politicians the worry of casualties provoking opposition at home.[29] Invaded countries have a harder time making a case to allies or the global community that they are suffering from the type of mass devastation that promoted interventions in other, more conventional wars.[30] War would look more and more like an internationalized police action, with suspect persons given a chance to defect or face mechanized detention.

The French philosopher Grégoire Chamayou is also skeptical, based on the history of technological hegemony in war settings. In his insightful book *Drone Theory,* he reminds readers of the slaughter of ten thousand Sudanese in 1898 by an Anglo-Egyptian force with machine guns that only suffered forty-eight casualties. Chamayou memorably brands the drone "the weapon of amnesiac postcolonial violence."[31] He also casts doubt on whether advances in robotics would actually result in the kind of precision that fans of killer robots promise.[32] Civilians are routinely killed by drones piloted by humans, and when it comes to imagining reform, it is difficult to imagine which is more chilling—automation of such force without proper identification of targets, or computing systems so intensely surveilling subject populations that they can assess the threat posed by each person within it (and liquidate accordingly). Even if we presume that the technology will become more targeted, there is still a massive leap in logic to assuming that commanders will buy it or use it or develop just principles of discrimination.

The convenient futurism of drone advocates is based on an idea of precision that Chamayou systematically dismantles. Drone advocates say the weapon is key to a more discriminating and humane warfare. But for Chamayou, "by ruling out the possibility of combat, the drone destroys the very possibility of any clear differentiation between combatants and noncombatants."[33] The assertion of the drone's "elimination of combat" may seem like hyperbole, but consider the situation on the ground in Yemen or Pakistani hinterlands: Is there really any serious resistance that the "militants" can sustain against a stream of hundreds or thousands of unmanned aerial vehicles from the US patrolling their skies? Such a controlled environment amounts to a fusion of war and policing absent many of the restrictions and safeguards critical to the legitimacy of each. No one should rush to legitimate it in military occupations.

UPPING THE ANTE IN GREAT-POWER RIVALRIES

Despite such powerful normative and ethical critiques, leading military theorists in great powers now appear to believe that they have little choice but to invest in the robotization of violence. Large militaries may decide to develop all manner of automated advantages. One strategy, for example, is to advance both defensively (by, say, investing in lightning-fast laser weaponry to neutralize enemy drones) and offensively (by building drones of one's own that can inflict brutal reprisal for any damage done).[34] Social theorist William Bogard calls this vision of permanent, ordered dominance a "military dream" of "the disappearing war, the war over before it is fought, all its uncertainties and indiscernible qualities . . . placed under rational control."[35] Gone are the days when generals might openly fantasize about utterly vanquishing an enemy. A model of "shock and awe" has become a matter of frightening an opponent into submission before force is even necessary.

If monetary and technological advantage were a clear public indicator of a military's capability to accomplish those ends, a Pax Robotica orchestrated by hegemons might dampen conflict. Complacency, however, seems inadvisable. Conflict expert Paul Scharre warns that automation creates a possibility of war robots making a "million mistakes a second."[36] Malfunctioning or hacked software could spark, rather than avoid, war. Even in the 1980s, white supremacist terrorists dreamed of provoking an American nuclear war with Russia as a way to end emerging multiculturalism and to install their own genocidal regime.[37] To such madmen, civilization-destroying nuclear winter was a small price to pay for racial purity. More prosaically, growing tensions in the Middle East, the South China Sea, Kashmir, and Ukraine give powers like the United States, Russia, India, Pakistan, and China ample opportunities to incrementally escalate the deployment of surveillance and armed drones on air, land, and sea.

The knife edge between offensive and defensive capabilities intensifies the danger. A logic of preemptive war plays into the paranoia of those afraid to lose their advantage.[38] Machines can be stealthier than humans. They can react faster. It takes a fighter pilot at least a third of a second to respond to attack; an automatic counter-fire system could observe, orient, decide, and act in under a millionth of a second.[39] Robotic defense systems could, in principle, hold *all* sensed assailants accountable for their actions. With a perfected automation of revenge, "anyone who would shoot at our forces would die. . . . [they] would have to pay with blood and guts every time they shoot at one of our folks," said a former member of the United States' Joint Forces Command in Iraq.[40] But hackers who spoofed the system could provoke a bloodbath and spiraling reprisals.[41]

"Eye for an eye" automation informs not only the military's approach to insurgencies, but even its response to far more formidable adversaries. In military strategy, the "doomsday device" has proven to be an object of both aspiration and ridicule. A ruthless power may try to kill the people in control of its rival's military. If the rival's forces were hard coded to respond with deadly force the moment any attack was detected, that first-strike strategy would be futile. Given devastating weaponry available globally, the logic of nuclear deterrence (mutually assured destruction) may apply a fortiori to autonomous systems with a dead man's switch.

Such a strategy is extremely dangerous. Automation can lead to disastrous results, causing the very danger it is meant to foreclose. In 1960, shortly before the US election, the United States' Ballistic Missile Early Warning System (based in Greenland) "detected" a Soviet missile launch, asserting "99.9% certainty" that one had occurred. It turned out that unusual light from the moon had set off the alarm. Fortunately, NATO stayed its plans for retaliation until after it learned about the mistake.

There are some striking real-world examples of out-of-control algorithms generating worrisome results that could in principle happen in military settings. For example, two booksellers on Amazon programmed bots that acted quite rationally individually—each would raise its price when it noticed the other raised its price.[42] When the two interacted together, they set off a feedback loop that ultimately priced at $2 million a book ordinarily worth about $30.[43] A computer glitch destroyed the firm Knight Capital, automatically generating tens of thousands of money-losing trades for the firm.[44] The flash crash of 2010 has been put down to unexpected interactions among far more complex trading algorithms.[45] The abstract patterns in any of these examples—of competition, one-upmanship, and fractally reticulated chains of reasoning leading to unexpected results—could also happen in weaponry that is increasingly computerized. Even if fail-safes and other safety mechanisms develop, they become yet more vulnerabilities for the enemy to hack. The risk would escalate as states devote ever greater resources to avoid domination by rivals.

BARRIERS TO BANS

Given these dangers, global leaders could try to ban certain functions or methods of killing outright, pursuant to principles of IHL. One foundational principle of IHL is "military necessity," which demands that commanders balance the practical requirements for success against a duty to "humanity."[46] Respect for the "dictates of public conscience" also inspires this body of law.[47] The vagueness of terms like "humanity" and "conscience" is a barrier to algorithmic logics, which demand clear values to trigger action.[48] Some studies show serious public concern about the rise of autonomous robots.[49] Human Rights Watch, the International Committee for Robot Arms Control, Pax Christi, Article 36, the Nobel Women's Initiative, and Pugwash have formed "a coalition to set up a civil society campaign for a new legally binding international treaty to ban the development, production and use of fully autonomous weapons."[50]

To understand the future of such arms control agreements, it is helpful to look at the past. The simplest of autonomous weapon systems is the booby trap set to capture an enemy. More lethal is a hidden bomb, programmed to detonate when some wire is tripped by the unsuspecting. The antipersonnel land mine, designed to kill or maim anyone who stepped on or near it, terrified combatants in World War I.[51] Cheap and easy to distribute, mines continued to be popular in smaller conflicts around the globe. By 1994, soldiers had lain 100 million land mines in sixty-two countries.[52]

Though small in size, the mines continued to devastate and intimidate populations for years after hostilities ceased. Mine casualties commonly lost at least one leg, sometimes two, and suffered collateral lacerations, infections, and trauma. The mines amounted to a public health crisis in some particularly hard-hit countries. In 1994, 1 in 236 Cambodians had lost at least one limb from mine detonations.[53]

By the mid-1990s, there was growing international consensus that land mines should be prohibited. The International Campaign to Ban Landmines (ICBL) pressured governments around the world to condemn them. The moral case for the ban was complex. The landmine is not nearly as deadly as many other arms. But unlike other applications of force, it could maim and kill noncombatants long after a battle was over. By 1997, when the ICBL (and its leader, Jody Williams) won a Nobel Peace Prize, dozens of countries signed on to an international treaty, with binding force, pledging not to manufacture, stockpile, or deploy such mines.

The United States demurred, and to this day it has not signed the anti-landmine weapons convention.[54] At the time of negotiations, both US and UK negotiators insisted that the real solution to the landmine problem was to assure that future mines would all automatically shut off after some fixed period of time—or had some remote control capabilities.[55] A device that could "phone home" could be switched off remotely once hostilities ceased.[56] It could, of course, be switched back on again, too.

America's technological solutionism did not appeal to many participants at the mine talks. Dozens of countries signed on to the Ottawa Accord by 1998.[57] More countries joined each year from 1998 to 2010, including major powers like China.[58] United States diplomats tended to defer to its military authorities on this issue. These military authorities, in turn, are notoriously skeptical of international arms control agreements. Their attitudes about such agreements are likely to accelerate the automation of warfare.

RESPONSIBILITY FOR WAR ROBOTICS

Instead of bans on killer robots, the US military establishment prefers regulation. Concerns about malfunctions, glitches, or other unintended consequences from automated weaponry have given rise to a measured discourse of reform around military robotics. For example, the New America Foundation's Peter W. Singer would allow a robot to make "autonomous use only of non-lethal weapons."[59] So an autonomous drone could patrol a desert and, say, stun a combatant or wrap him up in a net, but the "kill decision" would be left to humans alone. Under this rule, even if the combatant tried to destroy the drone, the drone could not destroy him.

Such rules would help transition war to peacekeeping and finally to a form of policing. Time between capture and kill decisions enables the due process necessary to assess guilt and set a punishment. Singer also emphasizes that "if a programmer gets an entire village blown up by mistake, he should be criminally prosecuted."[60] A strict liability standard would advance accountability for "weapons of math destruction" (data scientist Cathy O'Neil's memorable description of errant algorithms). Yet how likely is it, really, that the "mistaken" programmer would actually be punished? In 2015, the US military bombed a hospital manned by the Nobel Peace Prize winning international NGO

Medecins sans Frontieres. Even as the bombing was occurring, staff at the hospital frantically called their contacts in the US military to beg it to stop. Human beings have been directly responsible for drone attacks on hospitals, schools, wedding parties, and other inappropriate targets. It does not seem realistic to expect domestic or (what is left of) international legal systems to impose responsibility on programmers who cause similar, or even worse, carnage.

Whereas Arkin would code robots with algorithmic ethics, Singer builds on our centuries-long experience with regulating *persons*. But there is a sizable literature anticipating or hoping for fully autonomous robotics or software systems, unmoored from any person.[61] To ensure accountability for the deployment of "war algorithms," militaries would need to ensure that robots and algorithmic agents are traceable to and identified with their creators.[62]

Technology does allow such traceability. In the domestic context, scholars have proposed a "license plate for drones," to link any reckless or negligent flying to the drone's owner or controller.[63] Computer systems already try to solve or mitigate the "attribution problem" in cybersecurity (which occurs when someone attacks a system anonymously) by correlating signatures of action to known bad actors.[64] The fourth new law of robotics ("A robot must always indicate the identity of its creator, controller, or owner") should serve as a fundamental rule of warfare, and its violation punishable by severe sanctions.

MAKING FORCE COSTLIER

Developing global systems of attribution, as well an inspections regime to ensure weapons and computer systems are compatible with them, will be expensive. In ordinary economics, this would be a problem. But the topsy-turvy economics of technologies of destruction make this cost something of a benefit: it deters the accumulation of massive stockpiles of machines capable of inflicting harms on human beings.

There are many ways of making the deployment of force costlier, ranging from imposing legal conditions on the deployment of killer robots, to forcing human control of missions, to taxing such technology.

Such limitations, like strictures on surveillance, will be lamented by war futurists. They would prefer the freedom to develop and deploy autonomous forces. But inefficiency here is not just bureaucratic busy work—it reflects human values.[65] A robotized military or police state is an obvious threat to liberty. When it comes to state violence, the cost of meaningful human control over the deployment of force is also a benefit.[66] A "centaur" strategy, combining human expertise and AI, enhances both military effectiveness and ethical responsibility. That is not only because a pushbutton war may be premised on a mistake (a possibility all the more chilling in a world of fake news, irresponsible platforms, and advancing technologies of video and audio forgery).[67] Surveillance and control should be expensive endeavors, lest a garrison state cheaply monitor and control ever more of our lives.[68]

Global political economy matters, too. Investment in arms is a response to a perceived threat. In 2018, the United States raised its military spending to $649 billion. China spent the next most, at $250 billion. China's military spending has been growing for over two decades, but since 2013 at least, it has been kept at a consistent percentage (1.9 percent) of a growing GDP.[69] Saudi Arabia, India, France, and Russia were the next biggest spenders, buying between $60 and $70 billion of military equipment and services apiece. All these figures are misleadingly small, since they tend not to include expenditures on domestic security apparatuses, emergency appropriations, and long-term care for injured soldiers (or support for the families of those killed).

United States budget allocations for defense have tended to outpace the spending of the next seven countries combined. The massive military buildup will likely strike future historians as a profoundly misguided investment of national resources.[70] There is a grim irony in the world's leading military "superpower" being devastated by coronavirus in 2020, even after investing tens of billions of dollars in biosecurity, while so many other nations reduced cases and deaths far more deftly. As Dominique Moïsi, a political scientist at the *Institut Montaigne*, explained, "America prepared for the wrong kind of war. It prepared for a new 9 / 11, but instead a virus came."[71] Consumed by worries over phantom threats, American leaders invested trillions of dollars in arms while stinting on preventive measures that could have saved more lives than the country lost in the Vietnam, Iraq, and Afghanistan wars combined.

Joseph Stiglitz and Linda Bilmes have estimated that by 2008, Americans had devoted at least $3 trillion to war efforts in the Middle East. Bilmes updated the figure to $5 trillion by 2016. Numerous commentators have argued that despite some important victories against the Taliban (and the death of terrorists such as Osama bin Laden), this multi-trillion-dollar investment actually hurt the country's long term strategic interests.[72] The US military's hardware and software investment strategies have also been widely criticized by war analysts.[73]

Similarly, China analysts have chronicled the ways in which heavy-handed investments in Xinjiang (in both conventional security and AI-driven facial recognition and classification systems) have undermined the Chinese Communist Party's (CCP) larger aims at home and abroad. Ostensibly responding to a set of knife attacks by disgruntled Muslims, the Chinese state has now detained hundreds of thousands of persons there in reeducation camps. The CCP has also promoted high-tech modes of surveillance, loyalty assessment, and scoring. The extraordinary Hong Kong protests were rooted, at least in part, in a refusal to be dominated by what residents saw as an unaccountable and overwhelmingly powerful ruling class. Zeynep Tufekci reported that fear of a Xinjiang-like endgame—replete with ubiquitous security cameras (even inside homes) and constant assessments of loyalty to the regime—intensified the desperation of Hong Kong protesters. Their struggle helped flip the Taiwan election result in 2020. While Taiwan's Kuomintang was leading polls throughout 2019, they steadily lost support as the Hong Kong standoff dragged on. Pro-independence president Tsai Ing-wen handily won reelection, and will continue to cultivate public sentiment toward political separation and cultural distinctiveness.[74]

Opinions like those expressed in the last paragraph are unlikely to get widespread attention in China, thanks to the trends in public-sphere automation mentioned in Chapter 4. That AI-enhanced censorship may, in the short term, shore up the power of the regime. It is a weaponization of media intended to harden public sentiment against outside interference. In the long run, however, even non-democratic governments rely on some feedback mechanisms to determine what works and what does not, what conduces to legitimacy and what feeds the embers of smoldering resentments. As Henry Farrell has argued:

> A plausible feedback loop would see bias leading to error leading to further bias, and no ready ways to correct it. This of course, will be likely to be reinforced by the ordinary politics of authoritarianism, and the typical reluctance to correct leaders, even when their policies are leading to disaster. The flawed ideology of the leader (We must all study Comrade Xi thought to discover the truth!) and of the algorithm (machine learning is magic!) may reinforce each other in highly unfortunate ways.[75]

Again, the theme of strength-as-weakness recurs. It is the exceptional power of the CCP that enables it to so closely monitor and shape expression. But when such power stifles critical feedback, rigidity becomes brittleness and fragility.

This is not to say that China's great power rival, the United States, has much to offer in terms of contemporary governance models. Thanks in part to its own increasingly automated public sphere, the US is sliding toward post-democratic status. A fragmented media is unable to reflect and uphold basic social norms against destructive policies and for fair elections. The neurologist Robert A. Burton speculated that President Donald Trump could be modeled as "a black-box, first-generation artificial-intelligence president driven solely by self-selected data and widely fluctuating criteria of success."[76] Where Farrell saw Xi as blind to feedback, Burton characterizes Trump as utterly at its whim, mechanically testing cynical rhetorical appeals to determine "what works" as a way of shifting the blame for his manifold failures. Whatever the merits of a US-led global order in the past, the Trump Administration has both disdained it in the present and deeply tarnished its future prospects.[77]

The mere accumulation of power does not guarantee wise governance.[78] Multipolarity is not merely a likely outcome of great powers' governance failures, but also is necessary to protect a global community from the excessive power of hegemons that have become too tyrannical or chaotic to exercise power responsibly. There is a delicate balance between gradually reducing destructive capacity generally and ensuring that security alliances have enough firepower to deter aggression from greater powers and terrorism from non-state actors. Some states may well need to invest more in defense (including leading AI technology). But the greater a state's spending on weaponry currently is, the more its citizens should beware its contributing to the garrison state and arms race dynamics critiqued above.

There is a classic "guns and butter" trade-off posed by conventional public finance experts. Demands for more spending on human services are a drain on war-fighting posture. But such domestic spending may in fact save regimes from themselves by reducing the resources available to feed arms races and self-defeating shows of force. The more states are held responsible for delivering high and increasing standards of health, education, housing, and other necessities for their own citizenries, the less they will have to invest in the imperial projects made plausible by unfettered advance in military AI. Even when a smaller or poorer country understandably invests in lethal autonomous weapons to fend off depredations from great powers, it risks provoking similarly situated neighbors into investing to preserve their own relative position. The better option is to create strategic alliances to even the balance of forces.

Given the difficulty inherent in policing arms control agreements, pledges of self-restraint, even when entered into formal treaty agreements, are only part of a package of measures necessary to prevent AI-driven arms races. Just as crucial is a sustained social focus on projects as urgent and necessary as defense.

THE LOGIC OF COOPERATION

For a large body of researchers (often either directly or indirectly funded by militaries or their contractors), advances in autonomous robotic weaponry are inevitable. They invoke the logic of the slippery slope: there is a spectrum of human control over machinery and no clear principle to stop the further granting of autonomy to automated systems once initial steps have occurred. But technology need not develop in this direction. Forms of cooperation are as viable as ruthless competition. Norms have and will continue to deter creation of many possible weapons.

Given the myriad variables involved, the wicked problem of killer robots cannot be solved. But we can grow wiser about it via historical research into similarly intractable dilemmas. Scholars addressing the international political economy of nuclear proliferation have much to teach here. For example, Etel Solingen has explored why some nations and regions tipped toward investment in weapons of mass destruction, and others did not. Classic "realist" theories of international politics emphasize the power of arms races, rooting the decisions of countries to "go nuclear" in their sense of insecurity (which in turn is determined by the power of rivals and enemies). But these theories did not explain why so many countries failed to develop nuclear bombs, even when they faced security threats greater than those faced by old and new nuclear powers. Solingen argues that political leadership matters and that nations may rationally decide to focus on economic development (at home), as well as on commercial competition and cooperation (internationally), rather than on nuclearization.[79] A nuclear program is costly, both economically and reputationally.

Solingen closely examines the case of Taiwan, for example, which faced a uniquely difficult security situation throughout the mid-twentieth century, given China's treatment of it as a renegade province. Taiwan could have developed its nuclear energy program in a direction that would have created more weaponizable material. Instead, the leadership of the Kuomintang (long its ruling party) strategically opted to secure their legitimacy via economic growth and trade rather than doubling down on a confrontational stance with the mainland. Military investment declined as a percentage of a growing economy, and Taiwan's nuclear ambitions focused on peaceful uses.

Of course, Solingen's "focus of energies" theory is not the only rationale for this strategic posture. Taiwan's long-time leader, Chiang Kai-Shek, was an exile from the mainland and did not even want to contemplate bombing his countrymen. The United States had a strong commitment to nonproliferation and pressured Taiwan to preserve norms against the spread of nuclear weapons. Yet such factors also fit within Solingen's larger framework, as they underscore Taiwan's interconnectedness with great powers of the time.

In applying this political economy frame to LAWS, the key question is how to ensure not just norms and laws proscribing particularly destructive technology, but also the economic and reputational expense of pursuing them. Not just governments, but also firms, can play a constructive role here. OpenAI's reluctance in 2019 to release a speech-generating model offers one case in point. AI-driven text generation may not seem like much of a weapon. But once it is combined with automated creation of social media profiles (complete with deepfaked AVIs), bot speech is a perfect tool for authoritarian regimes to use to disrupt organic opinion formation online. Moreover, militaries can apply the technology to interfere with other nations' elections. Perhaps it is best not distributed at all.

"TECH WON'T BUILD IT" AND INTERNAL RESISTANCE TO THE COMMERCIALIZATION OF MILITARY AI

At some of the largest American AI firms, a growing number of software engineers are refusing to build killer robots—or even precursors of their development. Their resistance is part of a larger movement by professionals and union members to assert control over how they work and what they work on. Recent developments at Google show both the virtues and limits of this approach.

Government contracts offer lucrative and steady work to technology firms and indispensable expertise to militaries. Overwhelmed with footage accumulated throughout the world, the Pentagon contracted with Google to process images from drone surveillance. Drones may fly for thousands of hours over enemy territory. Humans can review the footage, but they may miss critical patterns that are a tell for terrorist activity. The promise of machine learning is to find such hidden features in recordings, and eventually to apply that analysis to real-time processing of images. Google offered a project named Maven to speed up the processing of video footage. With experience monitoring millions of hours of YouTube recordings per year for abusive images, copyright infringement, hate speech, and other issues, top Google managers aspired to apply these efficiencies to matters of life and death,

Those on the front lines of developing the tech, however, had a different moral calculus. Four thousand Google employees signed a short, sharply worded letter stating that "Google should not be in the business of war." The workers made both a business and a moral case for keeping well clear of military applications of AI technologies. Given the intimate reach of Google data collection into customers' lives, they felt that military applications would "irreparably damage" the corporate brand. Some firms, such as Palantir, Raytheon, and General Dynamics, were directly involved with military applications; others, such as Microsoft and Amazon, were indirectly involved. The Googlers said their firm should not be in either category. They closed their digital protest by asking their bosses to "draft, publicize, and enforce a clear policy stating that neither Google nor its contractors will ever build warfare technology." Though management axed the Maven project, at least ten Google employees quit in the wake of the controversy, worried that the firm would be able to avoid future ethically compromising work.[80]

While widely applauded in the press, the idealistic coders also faced critics. Some mocked what they saw as a belated show of concern. By 2013, the Snowden revelations had shown how useful big tech's data was to law enforcement and military interests. Google's former CEO, Eric Schmidt, had headed the Defense Innovation Advisory Board for years. Once famous for the motto "Don't be evil," Google has shed even that modest ethical standard.[81]

Other critics claimed that Google had not done nearly enough for the nation of its birth. There is an AI arms race, and if the United States can't tap its top tech companies, nations with pliant plutocrats will pull ahead. Sandro Gaycken, a senior adviser to NATO, commented, "These naive hippy developers from Silicon Valley don't understand—the CIA should force them."[82] Uncooperative firms risk America's falling behind China and even Russia in an AI arms race. Data integration at the scale of China's would likelier enable more rapid advances in monitoring and targeting individuals—capacities of great use in policing, epidemics, and war. Tight integration of the state and private actors may lead to more advances.[83] On the other hand, Western commentators may be exaggerating the threat.[84]

When Google announced Project Dragonfly, a search engine censored to meet Chinese government specifications, both realists and idealists were outraged. Why was China's government entitled to cooperation if the United States government was not? On the surface, the distinction is an obvious one: censored search was not going to kill anyone. However, thorny questions remained. What would happen to disappeared websites? Would Dragonfly inform on their authors? In an era of information warfare, tight monitoring and control of online life is a top priority for many states.[85] This authoritarian control can also cause major problems for the rest of the world—as in the case of SARS-CoV-2 (the virus responsible for COVID-19), a coronavirus that originated in China and that may have been contained more quickly, had Chinese officials not censored the physicians trying to warn about the spread of this dangerous new disease.

The Google activists' refusal to work on military tech may seem belated, or unpatriotic. However, the ethics of participation are contextual. The United States already has invested in extraordinary military advantages over other nations. Though its position may have slipped slightly since the 1990s, it is still capable of delivering apocalyptic violence to any country that attacks it. Projects like Maven may not advance that counteroffensive capability, but it does make a maintenance of the status quo—de facto occupation—far more plausible in many areas where the United States' presence is likely promoting more resentment than it is building goodwill.

PATHS TO COOPERATION

Guard labor, including preparations for war, is big business.[86] The rise of "terror capitalism" has turbocharged a thriving global trade in surveillance equipment. The United States is the leading exporter, but other powers are catching up. Chinese firms sell high-tech cameras to British security agencies and the Zimbabwean government.[87] Security specialists from around the world shop for wares at the Xinjiang Anti-terrorism

Technical Equipment Fair in China.[88] Russian military equipment shows up in Syria, Iran, and Mexico. Supposedly private firms stockpile data that is bought, hacked, or coopted by government "fusion centers" capable of integrating myriad information sources into pattern of life analysis of hundreds of thousands of individuals. Weaponry has always been big business, and an AI arms race promises profits to the tech savvy and politically well-connected.

Counseling against arms races may seem utterly unrealistic. Great powers are pouring resources into military applications of AI. All too many of their citizens either don't care or applaud any ostensible effort to bolster their security (whatever its cost effectiveness).[89] Yet that quiescent attitude may change over time as the domestic use of AI surveillance ratchets up, and that guard labor is increasingly identified with shadowy apparatuses of control, rather than democratically accountable local powers.

Military and policing AI is not used only, or even primarily, on foreign enemies. It has been repurposed to identify and fight enemies within. While nothing like the September 11 attacks have emerged over almost two decades in the United States, homeland security forces (especially at the state and local level) have quietly turned anti-terror tools against criminals, insurance frauds, and even protesters. China has hyped the threat of "Muslim terrorism" to round up a sizable percentage of its Uighurs into reeducation camps and to intimidate others with constant phone inspections and risk profiling. No one should be surprised if some Chinese equipment powers a US domestic intelligence apparatus, while massive US tech firms get coopted by the Chinese government into parallel surveillance projects.

The advance of AI guard labor is less a rivalry among great powers than a globe-encompassing project by corporate and government elites to maintain hegemony over restive populations. The powerful often have far more in common with one another than with the average citizens of their home countries. Jingoistic saber rattling paints a convenient veneer of conflict over a common interest in promoting the stable extraction of value from labor unable to demand a larger share of the economy's productive gains. To the extent that citizens of the United States, China, and Russia have some influence over the direction of their militaries, they should bear this new divide in mind. Just as we saw in the realm of policing, in war, the acquisition of spectacular new forms of AI-driven force is not a simple story of enhancing the government's power to impose justice. It is also a tool of oppression, using rapid technological advance to freeze into place an unjust status quo.[90]

Once deployed abroad in battles and occupations, military methods tend to find a way back to the home front. They are first deployed against unpopular or relatively powerless minorities and then spread to other groups. Despite *posse comitatus* principles, US Department of Homeland Security officials have gifted local police departments with tanks and armor. Sheriffs will be even more enthusiastic for AI-driven targeting and threat assessment. But just as we saw in the case of domestic policing and prisons, there are many ways to solve social problems. Not all require constant surveillance coupled with the mechanized threat of force.

Indeed, these may be the least effective way of ensuring security, either nationally or internationally. Drones have enabled the United States to maintain a presence in the Middle East and Central Asia for far longer than an occupying army would have persisted. The constant presence of a robotic watchman, capable of alerting soldiers to any threatening behavior, is a form of oppression. American defense forces may insist that threats from parts of Iraq and Pakistan are menacing enough to justify constant watchfulness, but they ignore the ways such hegemony can provoke the very anger it is meant to quell.

Grappling with the need for cooperation in an world that cannot be unilaterally mastered is a viable alternative to techno-capital's arms races. At present, though, the military-industrial complex is speeding us toward the development of "human-out-of-the-loop" drone swarms, ostensibly because only machines will be fast enough to anticipate the enemy's counterstrategies. This is a self-fulfilling prophecy, tending to spur an enemy's development of the very technology that supposedly justifies militarization of algorithms.[91] To break out of this self-destructive loop, we need to start grappling with the thought of public intellectuals who question the entire reformist discourse of imparting ethics to military robots. Rather than marginal improvements of a path to competition in war-fighting ability, we need a different path—to cooperation and peace, however fragile and difficult its achievement may be.[92]

In her book *How Everything Became War and the Military Became Everything,* former Pentagon official Rosa Brooks describes a growing realization among American defense experts that development, governance, and humanitarian aid are just as important to security as the projection of force, if not more so.[93] In an era of climate crisis, rapid responses to help (rather than control) those affected by disasters might slow or stop destabilization. In a similar vein, Chinese development programs, at their best, build infrastructure in partner nations that is positive sum, both enhancing the productive capacity of investees and sparking mutually beneficial trade.[94] A world with more real resources has less reason to pursue zero-sum wars. It will also be better equipped to fight natural enemies, such as novel coronaviruses. Had the United States invested a fraction of its military spending in public health capacities, it almost certainly would have avoided tens of thousands of deaths in 2020, as well as economically disastrous lockdowns.[95]

For this more expansive and humane mindset to prevail, its advocates must win a battle of ideas in their own countries about the proper role of government and the paradoxes of security. They must shift political aims away from domination and toward nurturance. Observing the growth of the US national security state—what he deems the Predator Empire—Ian G. R. Shaw asks, "Do we not see the ascent of control over compassion, security over support, capital over care, and war over welfare?"[96] Stopping that ascent should be the primary goal of contemporary AI and robotics policy. That will, in turn, require a new view of money, resources, and abundance. As the next chapter will show, the foundation of humane AI policy is a renewed political economy centered on human needs and opportunities.

7

· Rethinking the Political Economy of Automation

Each of the new laws of robotics has a cost. It is often cheaper to replace workers than to complement them with wage-increasing technology. Giving up on an arms race means risking the loss of profit and power. Delaying or avoiding the development of humanoid robots denies us potential mechanical companions. And ensuring that robots' behavior is attributable to some person or institution will impose complex record-keeping obligations.

But each of these costs has a flipside. What consumers spend, workers earn. When everyone abandons an arms race, all are better off. Delays of (or limitations on) robotic replacements of humans means there will be less competition from machines for human attention, affection, and resources. Attribution will deter accidents and help us hold wrongdoers accountable.

This doubleness reflects a fundamental tension in economics. The microeconomic perspective of individual businesses is to minimize labor costs. The macroeconomic perspective of society as a whole requires consumers to have some source of income—and for most of us, that is work. How society balances these rival demands of labor and capital is a subject of both politics and economics—also known as political economy. A political economy perspective opens possibilities that are now dismissed as impractical, or not even considered at all.[1]

Standard accounts of AI economics tend to pose a trade-off between regulation and innovation. Political economy rejects that idea. We are not trying to maximize the size of some imaginary economic "pie" and then redistribute the resulting bounty.[2] Rather, law can and should shape the type of AI we get. For example, law can limit or ban facial recognition systems that take thousands of pictures of schoolchildren each day. Law can require, promote, or subsidize robotic cleaners and testers to combat contagious diseases. Regulation of AI and robotics is not meant merely to rein in mad scientists or put some guard rails around rambunctious corporate titans. Rather, we are trying to preserve certain human values in health, education, journalism, policing, and many other fields. And we are trying to ensure that each of these fields can develop in response to diverse and local needs and priorities and not according to the dictates of firms now dominant in commercializing AI.[3]

That will require government funding. To be done well, the introduction of robotic systems and AI will often cost more, not less, than current approaches. The simple economic relationship between the first and third new laws of robotics should be clear: money not wasted in arms races for mere positional advantage can be devoted to complementary automation. But reallocating spending from guard labor to human services is just one part of the picture. We also need to create a more sustainable economy, with automation ordered primarily toward creating ample goods and services for all, while reducing that production's negative impact on the environment. That is where we should be doing more with less—not in the human services fields already damaged in so many places around the world by austerity.[4]

To clear the ground for my argument, I will need to debunk two narratives. One we might call an economist's nightmare, and the other, a radical's dream. The economist's nightmare is a societal diagnosis known as the "cost disease"—the fear that health and education spending parasitically drain resources from the rest of the economy. The radical dream is "full automation"—the hope that one day, everything will be done by machines. Economists tend to reject full automation as too utopian. Yet the dream and the nightmare are two sides of the same coin, a way of thinking that prioritizes efficiency in individual transactions over the harmonious arrangement of society as a whole.[5]

This chapter takes on both the cost-disease and full-automation narratives of the future of AI and robotics. Rather than allowing the largest technology firms to take over professions in the name of AI-driven efficiency gains, we must rebalance the economic playing field. The state must better protect the rights and prerogatives of workers and small to medium-sized enterprises, while limiting the power of vast technology firms. At present, too much law tilts capitalist economies toward substitutive automation. A revitalization of tax, competition, labor, and education policy would help right that balance, giving us all more of a stake in the future development of AI and robotics.

FROM JOB TRAINING TO INNOVATION GOVERNANCE

At present, market pressures push for fast and cheap automation. They model the world as an assembly line, with every job subject to being sliced and diced into smaller, routinized tasks. The business case for service robots is obvious—they do not need to be paid, to sleep, or to be motivated to stay on task. So long as there are tourists looking for a bargain or business travel departments looking to cut down expenses, hoteliers will want robot cleaners, receptionists, doormen, and concierges. All those jobs may become as obsolete as elevator operators.

The coronavirus pandemic has created even more pressures to reduce human interactions in service industries. Lockdowns of indefinite duration made the case for robotics better than any business guru. When warehouse operators, meat packers, and farmworkers fear catching a deadly virus at work, robotization of their roles may appear outright humanitarian (if paired with some plausible promise of basic income provision and future jobs). The moral balance in many other sectors of service automation changes as well. Being deemed an "essential worker" in the midst of a pandemic is a dubious honor. Even if a vaccine vanquishes the coronavirus strains that began ravaging the world in 2019, there is always a chance of another pandemic. This possibility, even if remote, adds force to the logic of workplace automation. A worse virus could decimate even essential services and supply chains, threatening social breakdown. The sudden materialization of the pandemic threat counsels in favor of accelerating robotics capable of assuring steady production and distribution of necessities.

But there will always be many social roles that demand a more human touch. We need to carefully draw a line between helpful, human-centered innovation, and mere exploitation of cost-pressures and crisis to promote premature automation. The story of robotics and AI is not simply one of inexorable technological advance. Wherever they are introduced, they will be greeted with both enthusiasm and trepidation, warmth and dismay. The job of policymakers is to decide how to reconcile these responses in a way that best combines respect for persons, recognition of resource constraints, and accountability. That will require immense and ongoing investment in education.

Astute universities already aim to make students "robot-proof," in the words of Northeastern University president Joseph Aoun.[6] Training in science, technology, engineering, and medicine (STEM) is in high demand. The humanities—ranging from literature to political science, history to philosophy—are also foundational to understanding the full stakes of automation, as Aoun wisely concludes. They must be revalued, not devalued. This is yet another way that AI properly implemented will lead to more jobs. Education becomes more labor-intensive when the most advanced modes of production become more complex.[7]

Economies do shift, and certain skills will become obsolete. The solution to this predictable problem is not to shorten secondary school or college, but rather to add opportunities for learning later on in life. Too many governments have only fitfully or half-heartedly funded programs designed to "retrain" workers. They should do more, moving beyond job training to more ambitious and fulfilling programs. A worker displaced by a machine deserves more than a few weeks at a coding boot camp. Opportunities to learn about why the displacement happened, how it might have been handled differently, and the larger context of economic change are vital, too. Such lessons used to be part of union schools, and they could be revived.

Even on traditional economic metrics, this strategy succeeds. Investing in education to keep up with the new opportunities that AI creates will promote productivity in at least three ways. First, it will serve to give more skills to those casualized or put out of work by advances in software, data analytics, and robotics. It will promote those advances, too, ensuring a more productive economy that can support people engaged in longer spells of education. Finally, jobs in research, teaching, and training should become more widespread, as new computational power reshapes work.

Artificial intelligence has required—and will continue to require—workers and managers to up their game, all while negotiating delicate and difficult disputes over the proper way to collect, store, and manage data. Advanced computation challenges us in our role as citizens, too. Two countries most resilient to an onslaught of online propaganda analyzed in Chapter 4 are France and Finland. Both have strong educational systems; Finland in particular has trained schoolchildren to identify the sources of messages online and to weigh ulterior motives and hidden agendas.[8] If democracy is to have a future, such investigation will need to become a far more common skill—as will basic values of inclusion, open-mindedness, and ability to reason about fairness and justice.

So many fields in social sciences and humanities have something to add to this process, as the very nature of expertise is iteratively shaped and reshaped over time. New fields (such as fairness, accountability, and transparency in machine learning) are developing in real time as universities meet both student and social demands. As computer scientists and operations researchers explore better ways to accomplish our goals, there will be a constant and necessary pushback from others capable of articulating the human values embedded in traditional approaches. When thought leaders in AI say humans need to be more emotionally "adaptable," to be understood by machines, thoughtful commentators can observe how such discipline distorts and degrades experience.[9] Apt emotional responses are more than mere sensation; they fuse knowledge and feeling in ways that disclose deeply troubling or worthwhile aspects of a situation. As we saw in earlier discussions of child and elder care, such responses embody something vital and human. Setting up institutions where we can respectfully explore such questions, rather than having them imposed by fiat, will be crucial.[10] At its best, the university provides these forums, educating students in sentiment and science, self-knowledge and world-discovery.

FROM SUBSIDY TO SUBSIDIARITY

Is it fair for the state to pay (either directly or indirectly) to help citizens understand how to best develop or limit technology? Can't the market supply such expertise? While a common objection in policy circles, this critique of subsidization fails on two levels. First, investments in education, research, and development, like genuine advances in health care, have enormous long-term benefits, sparking greater economic growth. Second, aside from any instrumental benefits, they are intrinsically worthwhile.

Individual market transactions are not always, or even usually, conducive to a broader social good. The crisis of global warming now shows that beyond dispute. Indeed, as the consequences of obesity and smoking demonstrate, such transactions are often problematic even for those engaged in them.[11] For investments with long-term social benefits, such as education, the market is even more ill poised to provide. Nor does the typical firm have the right incentives to improve workers' skills, given that they can easily decamp to rivals. Nor is advanced or specialized education something that the typical worker will (or should) risk much on without some kind of subsidy, particularly in an era of rapid technological and social change.

We need bold thinking about how to support and expand an education sector truly capable of preparing persons to take advantage of new opportunities in work (and leisure) presented by the digital revolution. Just as secondary education became publicly provided for free in the early nineteenth century, four years of postsecondary education should be a minimum entitlement for citizens. Some may complain that subsidies merely raise tuition costs. To the extent those costs are unjustified, price controls can also be imposed. But they should be used sparingly, since economic growth is enhanced both directly and indirectly by education.[12]

Economists of education are all too prone to miss the political dimensions of widespread diffusion of knowledge and other capabilities needed to govern and improve technology's deployment. To make democracy something more than an occasional visit to the ballot box, we need to give as many people as possible some role in shaping what

Andrew Feenberg calls the "technosystem"—the "second nature" of tools, media, and interfaces that so profoundly shape our lives. This is less an economic principle than an adaptation of an ideal of governance. Known as subsidiarity, this principle commends a devolution of responsibility to the most local entity capable of handling it well. A familiar form of subsidiarity is federalism—for example, the European Union or United States devolving responsibility for certain topics (such as the content of history education).

Maintaining human control over AI systems represents another form of subsidiarity, more functional than territorial. Imagine, for example, responsibility for a classroom. The most centralized approach would concentrate power over the exact material studied, discipline, recreation, and even bathroom breaks in a national authority. AI could empower such an authority with class monitoring cameras and software, baking into code the governance of conduct nationwide. More local control devolves some of that authority to superintendents, principals, and of course teachers. Preserving human control of AI systems will require similar delegation of responsibility across the economy. As one initiative of the Institute of Electrical and Electronics Engineers has stipulated, even where AI and robotics are "less expensive, more predictable, and easier to control than human employees, a core network of human employees should be maintained at every level of decision-making in order to ensure preservation of human autonomy, communication, and innovation."[13] At its best, this planning balances efficiency and democracy in the workplace, ensuring that well-trained professionals can evaluate where routinization via AI is going well, and where it can be improved.

It may seem odd to speak of democracy in the workplace—a conflation of political and economic terms. Yet "there is government whenever one person or group can tell others what they must do and when those others have to obey or suffer a penalty," as the legal theorist Robert Lee Hale observed.[14] Social organization demands some hierarchy, but its structure can be more or less open and contestable. AI should not entrench deep disparities in the power of workers, managers, and capital owners. Rather, it can help unions and worker associations bring more prerogatives of self-governance to the workplace. For example, algorithmic scheduling of workers need not be based merely on cost minimization, upending workers' lives with zero-hours contracts. Instead, organized laborers can demand that the relevant AI accommodate their needs dynamically, enabling exactly the type of family time, leisure, and educational opportunities that a more productive economy should be delivering to everyone.[15]

Democracy is about more than the obviously political realms of parties, elections, and legislatures. It should also extend in some form to the "private government" of workplaces now dominated by bosses.[16] Countries with robust industrial co-determination, like Germany, are better poised to apply AI and robotics fairly because they have already institutionalized representation of workers in the governance of firms.

The longer-term vision embedded in the first new law of robotics demands either professions or unions with some voice in the deployment of AI and robotics. In the short term, unions working in the public interest can blunt the most exploitative forms of AI management. In the long run, they may evolve into professions with distinctive, socially recognized prerogatives to apply their expertise (in situations ranging from data collection to evaluation). Teachers, doctors, and nurses assert themselves in this way, and their example is already influencing many other workplaces. Think, for instance, of Uber drivers capable of collectively bargaining for safety measures, or capable of disputing unfair evaluations of their work. In this vision, the drivers are not seen as mere data reserves, gradually paving the way for their own obsolescence. Rather, they are an integral part of a gradual social realization of a safer, faster, and more reliable transport infrastructure.[17]

Traditional economic approaches elide workplace politics by modeling employment as good-faith bargaining. That version of events is idealized at best and largely irrelevant absent robust social protections and laws protecting workers. Do you really dare bargain over the terms of employment? And if a boss deviates from these terms, what are the relative costs and benefits of insisting on your rights? Large employers staff full-time legal departments; how many workers enjoy the same? In short, many employment contracts are simply imposed, the way a city might impose ordinances on its residents—but without the forms of democratic control that legitimate how a public government functions.

Another strand of economic thought insists that we accept the unaccountable private government that employment-at-will creates, because this regime generates more social welfare than other ways of regulating the workplace. While such utilitarianism may be attractive as a bare-bones economic model of the cost of regulation, this generalization withers in the face of empirical evidence. It is by no means clear that the higher per capita GDP prevailing in the United States really indicates a better life for its citizens—especially its workers—than the more nuanced regimes of workplace governance prevailing in much of Europe.

Of course, there is no perfect solution here, and there will always be justified demands that those on the front lines of modern professional practice demonstrate their value. For example, authorities may need to rescue students from incompetent or martinetish teachers; neither market nor state can be expected to support forever a writer who is never read, or a musician who is never listened to. But the general structure of delegating responsibility for work to professionals is critical to avoiding a mental monoculture.[18]

THE INTRINSIC AND INSTRUMENTAL IN HIGHER EDUCATION

Aside from the subsidy question, there will also be predictable controversy over the content of higher education. Programs abound, ranging from the most pragmatically vocational to the highly impractical (if deeply rewarding). Given current trends in higher education policy, the greatest enemy here is premature specialization. College years should be roughly divided into intrinsic and instrumental studies. Instrumental studies are aimed at ensuring that a person has some marketable skills in a contemporary economy. These can range from coding to marketing, statistics to rhetoric. Intrinsically important subjects are the carriers of the values, tradition, heritage, and history that give us a sense of why our efforts matter.

Of course, there is no hard-and-fast division between the intrinsic and instrumental. Future lawyers can learn a great deal about close textual analysis by studying Renaissance poetry, and a yoga practice done at first only to improve one's health may eventually become a deeply resonant source of meaning. Nevertheless, given the rising popularity of business and similar majors at the undergraduate level, we are in grave danger of losing the "best that has been thought and said" without a stronger social commitment to invest in it. Given the enormous power of AI, we need to be in touch with deep human values as we deploy it.[19]

Some captains of computationally driven industries may balk at this suggestion. Anthony Levandowski, an engineer behind some key advances in self-driving cars, once commented that "the only thing that matters is the future. I don't even know why we study history. It's entertaining, I guess—the dinosaurs and the Neanderthals and the Industrial Revolution, and stuff like that. But what already happened doesn't really matter. You don't need to know that history to build on what they made. In technology, all that matters is tomorrow."[20] This dismissal of the past is troubling. Those at the commanding heights of transport technology need to understand why, for example, car culture may have been an enormous mistake. A deeply learned and humane book like Ben Green's *The Smart Enough City* tells that story with authority and should be read by anyone rushing to double down on automobile-bound individualism when collective transport solutions are so critical to reducing greenhouse gases.[21]

Economists of higher education do not usually speak in this register of values. All too much of the policy space in this realm has been dominated by projections of the "degree premium"—essentially, increase in actual earnings attributable to education, minus its costs. As with any misguided project of quantification, these cost-benefit analyses do not take into account all relevant data, such as the social value of the work graduates do. By crudely shaping the education sector itself toward becoming a purely instrumental endeavor in "workforce preparation," they help extinguish the very values necessary to recognize economism's shortcomings.[22]

RETOOLING TAX POLICY IN AN AGE OF AUTOMATION

To ensure robust job opportunities, there is a kaleidoscopic array of policy options, all capable of subtle adjustment to balance the many social objectives in play. For example, an earned income tax credit (EITC) is designed to induce workers to do labor they ordinarily would not do (or to reward work that is chronically underpaid). In general, income taxes deter labor at the margin, since every additional dollar paid by an employer results in less than a dollar received by a worker. I may work to make an extra dollar, but if the tax rate is 25 percent, I only get 75 cents. With an EITC, those at lower income levels actually receive a bonus from government for working rather than paying a tax on income. So, for example, a taxi driver may take home $13,000, despite only earning $10,000 from fares, because the transfer payments effectively create a negative income tax of 30 percent. While stingy in the United States, the EITC could be boosted to improve incomes for those in sectors where there is less work to do because of automation. And the program seems particularly apt for consideration in places like the United Kingdom and Germany, where the marginal tax rate for people coming off unemployment can hit 80 percent.[23]

Of course, programs such as the EITC have many shortcomings. They only offer help to people with some level of income. For those stuck in sectors where AI has effectively taken *all* work, the EITC is a cold comfort. Subsidizing low-wage work also means that we will see more of it. In other words, firms that could be innovating to increase the productivity of workers may instead be lulled into keeping suboptimal labor arrangements in place. There is less reason, for example, for a chain of stores to invest in automated cleaning equipment to replace janitors (or even make them 10 percent more efficient) if state subsidies effectively give labor a massive cost advantage relative to machines. Of course, programs like the EITC may merely be making up for the unfair treatment of wages relative to capital income in many tax systems. Nevertheless, we need to avoid deterring technological advance in areas where human governance and insight are not contributing to process and quality improvements. And we must also avoid supporting "lowest common denominator" employers who have browbeaten their workers into ever-paltrier wages.

These predictable shortcomings of the EITC have fueled widespread interest in another, simpler approach: a universal basic income (UBI) for all persons, whether or not they work. Trialed repeatedly and exhaustively defended and developed by philosophers such as Philippe van Parijs and Yannick Vanderborght, UBI has enjoyed renewed relevance in an era of automation. Four distinct framings justify it. One is purely humanitarian: everyone deserves some basic subsistence, regardless of their contributions to society. A second is Keynesian: a UBI would raise economic activity because many of its beneficiaries have a high marginal propensity to consume (or, in less elevated language, money is like manure: far more fertile and generative when spread around than when piled into a heap).

A third conception of UBI comes from philosopher and sociologist Hartmut Rosa, who sees it as a material foundation for "resonance."[24] For Rosa, modernity's tendency toward alienation—a sense of meaninglessness and blankness—necessitates a renewed investment in the resonant—those sources of meaning and value that we so often find in nature, art, history, play, our family, and our friends. By taking the worst forms of poverty off the table of vocational calculations, the UBI is supposed to give us the space and time to engage in work we truly care about—or to not work at all.

Rosa's vision is inspiring. But worries about opportunists dropping out of the labor market (while others toil in socially necessary roles) fuel popular opposition to UBI. A fourth normative justification for UBI tries to meet that challenge by insisting that nearly all of us (and our ancestors) have had some role in building the material foundations for whatever level of prosperity we all now enjoy. For example, merely using Google or Facebook tends to improve those services' algorithms. Comments, likes, and other reactions refine algorithms even further. Performers work for an audience, but the audience also works for performers, offering moment-by-moment judgments on the value of the entertainment provided by either paying attention, succumbing to distraction, or even frowning, huffing, or showing other signs of impatience. This may seem like an over-expansive definition of labor. But part of valuing and maintaining the human is a willingness to recognize the importance of social institutions that give meaning and purpose to individuals. Rather than condemning UBI for paying people to be frivolous, we may well find that making it easier for citizens to engage with culture is a commendable aspiration.

The universal contribution to the productivity of the economy demands something like a universal destination for human goods.[25] When millions of workers are displaced by robots and AI, the technology will have been trained on data the workers created in the course of doing their jobs. In a different regime of intellectual property, they might be able to demand royalties for such work, perhaps even setting up streams of rents for generations to come. Universal basic income would compensate workers generally for such unsung contributions to productivity.[26]

Where would the money come from? One idea is to tax robots and AI to level the fiscal playing field. There is an asymmetry at the heart of tax codes that "aggressively subsidize . . . the use of equipment (for example, via various tax credits and accelerated amortization) and tax . . . the employment of labor (for example, via payroll taxes)."[27] We have already seen many instances where the premature or misguided application of artificial intelligence instead of intelligence augmentation has caused both direct problems (of bias and inaccuracy) and indirect ones (ranging from a loss of dignity to a de-skilling of professionals who ought to be guiding, rather than replaced by, AI and software). Yet there are also vast swaths of the economy—jobs in cleaning, infrastructure, transport, clean energy, logistics, and much more—where we desperately need more and faster application of robots and AI. So it would seem unwise to tax robotization as a whole, particularly when these sectors need to rapidly advance to give everyone some semblance of the capabilities now taken for granted by the globe's wealthiest quintile.

Another, better source of revenue for a UBI would be taxes on wealth or on high income levels. In some countries, inequality is at such extreme levels that these taxes could raise enormous sums in aggregate. For example, US senator Elizabeth Warren proposed a US wealth tax targeted at the 75,000 US households with over $50 million in assets, which would raise $275 billion a year for ten years.[28] Senator Bernie Sanders would raise $435 billion annually from 180,000 households.[29] Each of these plans could fund important social programs. By contrast, if the Sanders plan's $435 billion were simply divided up among all 333 million Americans, they would receive an annual basic income of about $1,300. This would certainly improve the finances of many households—

particularly those with children—but would not be a serious candidate for actually replacing occupational income.

Of course, theorists can and have developed much more ambitious plans for a UBI, based on broader and deeper tax levels. For example, Karl Widerquist has proposed a UBI to ensure no household in the United States has less than $20,000 in income per year.[30] Starting from a relatively low threshold, it would tax income at a 50 percent marginal rate. That would, on net, subsidize households making between zero and $50,000 in income with an amount ranging from roughly $20,000 down to $2,000. By the time a household makes $55,000 per year, it would be paying more into the UBI than it was receiving.

Of course, there is a world of difference between a UBI paid for by a wide range of workers and one funded by taxes on wealthy investors. There are countless possible adjustments to the tax targeting that one could do in order to land somewhere between the $1,300 and $20,000 per year figures. The more generous a UBI is, the more likely it is to provoke political resistance to the taxation necessary to fund it. Advocates of a UBI could greatly strengthen their case by focusing their first demands for redistribution on fairly taxing the wealthy, and then gradually expanding their proposed tax base.[31]

FROM UNIVERSAL BASIC INCOME TO JOB GUARANTEE

Funding issues may not be the most significant challenge for a UBI. Whether framed as humanitarian relief, Keynesian intervention to stimulate the economy, or just compensation, UBI runs into practical difficulties quickly. Landlords will not ignore a new subvention; they may simply raise rents to capture it. Indeed, platform capitalists like AirBnB are even now accelerating the monetization of "spare rooms," eerily reminiscent of a Tory policy to impose a "bedroom tax" on benefits recipients in the United Kingdom whom the state deemed to have too much space in their homes. Other powerful players in the contemporary economy are also likely to raise prices. This is a particular danger for the United States, where, as French economist Thomas Philippon has shown, key sectors are dominated by oligopolists. If it is poorly planned, UBI simply feeds into inflation.

Another justifiable worry centers on the UBI as an unraveler of the state. The concept gained support on the right from libertarians who wanted to offer citizens a simple bargain: taxation would continue, but its proceeds would be distributed to everyone equally, rather than supporting health insurance, education, post offices, or similar state-provided services. Citizens could use those payments to buy what was once provided by the state. Some Silicon Valley futurists have seized on UBI as a substitute for state-provided services. Simply giving everyone money to purchase health insurance or schooling would, by this line of reasoning, enable optimal freedom in consumption choices. Some might choose very cheap health insurance and use the rest of the money for vacations or a nicer car; some families might opt for cheaper cyber-schooling at home and use the extra money to buy luxuries. It is not difficult to see the limits of this strategy; children in particular deserve some protection from the bad consumption choices of their parents. Those with cheap insurance force a terrible choice on hospital personnel once they are injured: treat with little to no guarantee of compensation, or let the person die (with attendant moral opprobrium). This is one reason why most supporters of a UBI frame it as an explicit add-on to extant state services rather than a replacement of them.

Even when those add-on assurances are made, however, there is a risk of payments like UBI subtly unraveling support for public purposes. For example, over the past few decades, Alaska has accumulated a "permanent fund" via royalties on resource extraction in the state. The fund has made payments of about $1,500 (in 2019 dollars) per resident for decades. A Republican governor of the state wanted to raise the payment and, to that end, pushed massive cuts in the state's education budget (targeted at universities).[32] The more states get in the business of a direct cash transfer out of general revenues, the more pressure there will be on legislators to cut services to pay for a more robust UBI. While many advocates imagine UBI funding primarily from taxes on the well off, the rich are very well organized politically. The easiest source of funds is cutting funding from investment in future productive capacity (like education or infrastructure). For example, rather than maintain tax rates on the rich, the United States cut funds from pandemic preparedness (among other programs), only to find itself utterly overwhelmed when COVID-19 struck. Indeed, the history of US decline over the past half-century could be written as a story of the systematic transfer of once-public funding for infrastructure, education, and health care back to private individuals and corporations.

Even the most sophisticated theories of a UBI tend to have two common shortcomings. First, they tend to model the economy as a machine distinct from persons and society, generating bounty (rather like a goose laying golden eggs). The reductio ad absurdum here is so-called fully automated luxury communism, with productive capacities needing little to no intervention from citizens. Second, they assume that individual consumption decisions are optimal allocators of resources. But there are many ways for such choices to go wrong— including by corroding the economic "machine" that the UBI vision relies on.

Imagine, for example, two societies that each implement a UBI. In one, recipients put the money toward larger automobiles with higher gasoline consumption. In another, the preferred spending is on rooftop solar panels that generate electricity for hybrid vehicles. The solar nation is not only helping the environment; it is also becoming more resilient, more capable of weathering volatile fossil fuel prices or the type of carbon taxes that should become a cornerstone of more sustainable global governance. As long as low-hanging fruit like solarization beckon, the wisdom of AI policy based on direct subventions to individuals will be questionable.

Critics of UBI have also pursued a broader vision for social purposes to be funded by an automation dividend. Rather than income, one line of research proposes government provision of universal basic services (UBS), which could include health care, housing, electricity, broadband access, and more. Rather than guarantee income, UBS would guarantee jobs in those necessary fields. Recent work on UBS should be a focus of debate on the political economy of automation, because it helps us understand the deepest needs in contemporary societies. To the extent AI and robotics innovation can be directed, consistent with the four laws of robotics mentioned above, it should be directed there.[33]

Given the coronavirus-sparked economic crisis of the early 2020s, there will be ample opportunity to guarantee jobs in such fields. The

US New Deal offers one model; its Civilian Conservation Corps and Works Progress Administration rapidly employed the jobless during the Great Depression. The economist Pavlina Tcherneva has promoted a more bottom-up approach based on proposals by community groups and nonprofits.[34] The macroeconomics of assistance are key here. The goal is not simply to replace the status quo but to ensure employment policy in the wake of AI-driven job loss addresses the triple (and mutually reinforcing) threats of climate change, ill health, and gross inequality. These challenges must be at the center of any sustainable "future of work" policy.

QUESTIONING THE COST DISEASE

The idea of government guaranteeing jobs in human services will be sure to meet resistance among conventional economists. Their elevation of a policy preference for austerity into a cornerstone of public accounting began decades ago. Deemed the "cost disease" by economists William Baumol and William Bowen, this approach divides the economy into two sectors.[35] In the so-called progressive sectors, like manufacturing and agriculture, prices fall even as quality rises for consumers. Today's farmers can use advanced technology (including robotics) to generate far more grain, at higher quality, than their forbears.[36] The same holds true in manufacturing; Ford's River Rouge factory in Dearborn, Michigan, once employed twenty times more workers at a time when it produced fewer cars than it does today. In these "progressive" sectors, employment falls over time as machines do more work.[37]

By contrast, in what Baumol and Bowen called the "stagnant" sectors of arts, health, and education, prices either stay the same or rise. It took four people to perform a string quartet in 1790; the same number are needed today. A lecture course in 2020 may share the same basic pattern as one in 1820 or 1620. For cost-disease scholars, the continuity of professors' lecturing, or doctors' physical exams, is suspicious. Why isn't the school as revolutionized as the farm, the students as robust as new genetically modified crops? Where are health care's assembly lines, its interchangeable parts and standardized interventions?

It is common to witness waste or inefficiency in any large institution, so challenges like these are superficially attractive. But the cost-disease diagnosis is little more than an ensemble of bad metaphors and dubious assumptions. Automation may reign supreme in the making of things or the transporting of them to places, but students and the sick are not products. Determining the bounds of the curable, or even the limits of treatment in difficult cases, is a dynamic and dialogic process. High-quality end-of-life care demands searching inquiries about a patient's goals and values. At the beginning of adulthood, finding a field worth learning and developing a vocation out of it is also an exceptionally difficult task, too idiosyncratic to be standardized into forms. Diverse communities of teaching, research, and learning must continue to aid individuals as they face these difficult decisions (and practices).

Global labor platforms to find the lowest-cost producers should not colonize service sectors. Instead, we need to re-commit to local governance by flexible and empowered professionals, both to assure good service now and ongoing advances in AI. Professionals perform a distinctive form of work, and they owe certain ethical duties to their clients. For example, fiduciary duties require looking out for the best interests of clients, rather than simply advancing one's own commercial interests.[38] Psychiatrists, for instance, should treat only those who actually need their help, even if they could make far more money by recommending weekly sessions, in perpetuity, for everyone. Contrast that duty with, say, the minimal commercial ethics of salespeople or marketers, who have no obligation to inquire whether the customers they solicit need (or can even afford) their wares.[39] Health apps, too, are often unbounded by professional norms, which is one reason why so many mental health professionals are now raising concerns about them.

Professionals also enjoy forms of autonomy rare in other fields. For example, the courts tend to defer to the decisions of licensing boards as to who is fit to practice medicine and who is not. They largely stay out of school decisions as to whether to suspend students accused of disruptive behavior.[40] Shared governance, including tenure protections, is a hallmark of respectable academic institutions. Respect for educator autonomy stems from a trust that educators will not abuse such a privilege—for example, that they will not allow parents to buy better grades for their children, or sell off their research time to the highest bidder. We have not yet perfected the kinds of institutions necessary to ensure such responsibility among those marketing AI and robotics.

Professionals have been granted some degree of autonomy because they are charged with protecting distinct, non-economic values that society has deemed desirable. Their labor, in turn, reflects, reproduces, and is enriched by those values. Knowledge, skill, and ethics are inextricably intertwined.[41] We cannot simply make a machine to "get the job done" in most complex human services fields, because frequently, task definition is a critical part of the job itself.[42] In the face of well-hyped automation, professionals ought to reaffirm their own norms, highlight the importance of tacit skills and knowledge, and work to extend their status to other workers who are capable and willing to codirect the development of AI and robotics in their own field.

The sociologist Harold Wilensky once observed that "many occupations engage in heroic struggles for professional identification; few make the grade."[43] But if we are to maintain a democratic society rather than give ourselves over to the rise of the robots—or to those who bid them to rise—then we must spread the status and autonomy now enjoyed by professionals in fields like law and medicine to information retrieval, dispute resolution, elder care, marketing, planning, designing, and many other fields. Imagine a labor movement built on solidarity between workers who specialize in non-routine tasks. If they succeed in uniting, they might project a vision of labor far more concrete and realistic than the feudal futurism of techno-utopians. They would foster automation that complements extant skills and labor, rather than accelerates a cheaper, faster, and more catastrophically unequal version of the present.

In an era of automation, our primary problem is not how to make labor-intensive services cheaper.[44] Pervasive wage cuts play into the liquidationist illogic of the Ouroboros (a mythological self-devouring snake): people who benefit in their role as consumers end up losing out as producers, and also as citizens.[45] For example, a retailer may at first be thrilled if AI-powered cyber-schools break a teacher's union; cheaper public schools presumably mean lower taxes. But he may then find that teachers no longer have the money to shop at his store. This is not a new problem; Keynes recognized it in the early twentieth century as a "paradox of thrift," threatening to condemn struggling economies to a downward spiral of deflation: lower prices, lower pay, and consequent unwillingness of consumers to spend even on low-priced goods, leading to even cheaper prices and a reinforcement of the same disastrous dynamic. From the United States in the 1930s to the Japan of the 2000s, this paradox of thrift has afflicted real economies. And it is clearly one of the biggest problems arising out of a future of mass automation, where tens of millions of workers may be replaced by machines.

Strangely, the paradox of thrift barely registers in contemporary economic thinking about automation and labor, when it appears at all. Skill building is still the preferred policy prescription, despite its obvious inability to address the many challenges posed by automation.[46] Even gatherings devoted to the problem of robots taking jobs tend to ignore troubling contradictions in current economic policy.

For example, technology advisors in the Obama administration hosted high-level workshops on the economic consequences of AI and robotics.[47] A common theme among speakers was an insistence that automation is far from matching human capabilities in many areas, and that automation, done well, would require intense investment in fields such as health care and education.[48] Yet at the same time, neoliberal progressives were consistently pushing policies to disrupt health care and education, reducing the flow of funds into these sectors, accelerating degree programs, and tightening the budgets of hospitals and other caregiving institutions. Some even explicitly connected this agenda to the alleged need for more military spending (feeding into the arms races mentioned above). Centrists around the world have adopted these policies, as have more conservative parties. In an economy in which technological unemployment is a major threat, one would think they would be thankful for new positions at hospitals, nursing homes, schools, and universities. But too many economists remain conflicted, anxious about maintaining some arbitrary caps on spending in certain sectors while oblivious to unnecessary cost growth in others.[49]

There is nothing intrinsically more rewarding about working an assembly line than providing, say, physical therapy or companionship to the injured, disabled, and elderly. Political theorist Alyssa Battistoni has argued that there is also an ecological case for shifting purchasing power away from manufacturing and toward services: pink collar jobs are green, in the sense that they use less carbon and other resources than physical objects.[50]

A common rejoinder is that to be internationally competitive, economies must focus productive capacity on exportable goods and services.[51] But countries that excel in education attract millions of students from abroad, and medical tourism is also a growing field. Human services are not a second-class occupation, subordinate to an economy's ultimate end of generating "things," such as cars, weapons, houses, and computers. Moreover, the cost-disease thesis itself indicates how futile such an approach is likely to be. The point of progress in these so-called progressive sectors is to gradually cheapen what they make. To cap health and education spending as some percentage of that declining amount of funds is a prelude to a deflationary spiral, not prosperity.

Wealthy economies don't thrive simply because they have a well-educated workforce and productive technologies. They also depend on a certain forward momentum in investment and saving. The real danger of mass, unplanned automation is a destruction of that momentum, as fearful consumers undercut firms and governments, and ultimately themselves, by saving as much money as conservatively as possible.

A NATURAL ECONOMIC TRANSITION

Aging populations will create extraordinary demand for care.[52] This is hard work that society should fairly compensate.[53] At the same time, automation threatens to replace millions of extant blue collar jobs, especially among those making less than $20 an hour.[54] This situation suggests a natural match between distressed or underemployed workers (now being replaced by self-driving cars, self-check-out kiosks, and other robotics) and emerging jobs in the health sector (for home health aides, health coaches, hospice nurses, and many other positions). Those jobs in health care can only emerge if policymakers value the hard work now done (and remaining to be done) for the sick and disabled.[55]

The need for more caregivers—and robotics to assist them—is obvious. Families endure stress and strain when they cannot find skilled, professional caregivers.[56] Helping them might increase health care costs, but it could lead to net economic gains overall, particularly for women, who disproportionately shoulder the burden of unpaid caregiving. A policy debate relentlessly focused on reducing health costs misses the opportunities that care creates. As Baumol observed in 2012:

> If improvements to health care . . . are hindered by the illusion that we cannot afford them, we will all be forced to suffer from self-inflicted wounds. The very definition of rising productivity ensures that the future will offer us a cornucopia of desirable services and abundant products. The main threat to this happy prospect is the illusion that society cannot afford them, with resulting political developments—such as calls for reduced governmental revenues entwined with demands that budgets always be in balance—that deny these benefits to our descendants.[57]

Technocratic health economists, myopically focused on deficit projections, contribute to the illusion that we cannot afford progress by smuggling an ideology of austerity into ostensibly neutral discussions about the size of the health care sector. Indeed, as the COVID-19 crisis demonstrates, such investment may be necessary merely to support the basic functions of the economy. Preparing for other low-probability but high-impact disasters should continue to be a source of jobs requiring human insight, cooperation, and judgment.

Steady investment in such services will promote productive options for those displaced by machines. After slashing cuts to social provision in the 1990s, China has steadily invested a rising percentage of a growing GDP in the health sector. If successful, such a transition may mark a triumph of "crowd-in" of Keynesian expansion, rather than "crowd-out" of other expenditures by health spending. At present, savings rates are unusually high because so many workers worry that their (or their parents' or children's) health expenses will not be affordable. A state that provides health security to all liberates them to spend more on consumption, rebalancing an economy tilted toward excess saving.

Promoting inexpensive, mechanized versions of services as replacements for professionals may erode the social processes needed to ensure its future progress. And as we saw in earlier discussions of education and health care, it is not enough to set up AI simply to measure outcomes and maximize positive ones. Professionals are needed at every step of the process to gather data, explain its implications, and critically evaluate the strengths and limits of new tools and decision support.

For example, consider a National Academy of Sciences report recommending college degree attainment by caregivers for young children.[58] Between the ages of two and eight, children experience extraordinary social and intellectual development—and vulnerability.[59] Understanding the latest in psychology, brain science, and related disciplines could help childcare workers do their job much more effectively and help them gain the professional status they deserve. They should also learn about a large and growing literature on AI in education, computerized personalization of lessons, and complementary robotics. An intensive postsecondary degree program, plus ongoing continuing education, will be essential for that. Yet far too many unimaginative commentators blasted the guidelines as an expensive burden on labor.

In medicine, journalism, the arts, education, and many other fields, services rendered invariably reflect cultural values and aspirations. They are fundamentally different from manufacturers or logistics. Automation may reign supreme in the making of things or the transporting of them to places, but readers, students, and the sick are not products. Determining the bounds of the curable, or even the limits of treatment in difficult cases, is a dynamic and dialogic process. However tempting fantasies of optimization-by-automation may become, persons should not be subject to them.

HOW TO PAY FOR IT: THE RENEWAL OF PUBLIC FINANCE

Among too many leading pundits and politicians, there is a common understanding that government spending is like that of a household. This conventional wisdom entails several commitments. First, they believe governments would ideally save more than they spend. Second, they assume a realm of economic activity that preexists government. To fund its activities, government must tax the private income and capital generated by markets. Third, they predict disaster for governments whose money creation is unmoored from the constraint of private demand for their debts. What the gold standard once did, the golden handcuffs of the bond market now achieve: ensuring that states only issue a stable amount of currency.

Another approach, called modern monetary theory (MMT), aims to upend this common sense of money at every turn by emphasizing the power of governments relative to other institutions.[60] A household cannot tax thousands of other households in order to support itself; government can. A household cannot print widely accepted money to pay for its expenses; governments that issue their own currencies can. Sovereign debt is not necessary for a sovereign currency issuer, but even if debt were issued, the issuer need not default—it can print the money necessary to pay off bondholders.[61]

Yes, the sovereign risks inflation if it creates too much of its own currency. But the hope behind MMT is that wise spending will increase the productive capacity of the economy, sparking innovation to assure more efficient use of resources and mobilizing now-unemployed labor. Energy in a country full of solar panels will be cheaper than is one without them, *ceteris paribus*. Better transit systems will cheapen goods by reducing the transport increment of their cost to consumers. Housing insulation projects reduce heating and cooling costs. Education raises wages; preventive care reduces medical bills. The list goes on; there are endless projects capable of paying ample dividends to recover the cost of investment in them. What is odd is that this idea of investment seems to be entirely associated with the private sector. As economist Mariana Mazzucato has argued, the state must share in this prestige as well, especially for long-term projects like decarbonization and adaptation to AI, which a short-termist private sector will never address adequately.[62]

Of course, not all MMT-funded projects will bear fruit. Even those that do may temporarily cause shortages of resources. This is the foundation for the legitimate fear that increasing the money supply will cause inflation. But there are many ways for government to handle such problems. It can tax money out of the economy, smothering the tinder of inflation. That taxation may be painful, but it is to be meted out, by and large, to those most able to spark inflation (those with the most money) or those most responsible for it. Sector-specific strategies also make sense. For example, requiring larger down payments for mortgages or taxing home sales can prick incipient housing bubbles. Modern monetary theory also has a broader theory of the particularity of inflation—that is, that at least in early stages, inflation is sparked in particular goods and services in particular markets. Focusing on those areas specifically can help avoid the disruptive effects of classic anti-inflation interventions, like the Volcker shock. Governments can also reduce inflation by altering rules governing the creation of private credit (bank money), such as bank reserve requirements.

In other words, sovereign currency issuers face an inflation constraint in their spending, not a debt constraint. Investing in AI and robotics that cheapen products, while enhancing the value of labor in professional (or professionalizing) service sectors, could be mildly inflationary on balance. But this is no reason to succumb to the deflationary spiral promised by unmanaged automation (or simultaneous shocks to demand and supply, like the COVID-19 pandemic). As soon as such an increase in price levels becomes troubling, policymakers can intervene to direct targeted regulation to reduce its impact.

This balance between hard and easy money, between the necessities of scarcity and prevalence, has been the subject of countless battles over currency. Increases in the money supply have helped unleash contemporary economic growth. But there are always voices of retrenchment and austerity. Like a gold standard, Bitcoin is deflationary by design. As currency becomes harder to mine and thus rarer and dearer, people who hold on to it are supposed to get richer relative to spenders. That is great for those who got in early hoarding gold or cryptocurrency, but its logic is disastrous for economic growth if it becomes a pervasively held ideal. As Keynes explained, the individually rational (saving money) can be collectively harmful (contracting economic exchange that gradually puts people out of work).

As that Keynesian parable suggests, aspects of MMT are decades old—and some of its oft-repeated insights are older than that. Keynesianism took root amid a global depression, when unemployment sparked enormous personal suffering and political upheaval. Its revival now is unsurprising, given the prospect of a pandemic-induced global depression. A job guarantee is meant not merely to put underutilized labor to work, but also to set a floor of compensation for workers that the (rest of the) private sector must best in order to attract workers.

Modern monetary theory also makes some important deviations from Keynesian orthodoxy. One reason Keynes became a foundation for modern macroeconomics is the studied neutrality of his doctrine. Keynes could joke that it mattered not whether a state endeavoring to put its citizens back to work spent money on burying and exhuming bottles or building a new pyramid of Cheops. In our time, the disastrously pollutive effects of much consumption are well known. So the political face of MMT in our time is not simply an argument for a "people's quantitative easing" or a universal basic income (both of which would undoubtedly reduce unemployment to some degree). Rather, it is a Green New Deal, an investment in the types of productive capacity that can decarbonize (or at least not contribute to the carbonization of) the atmosphere.[63] This substantive emphasis is a major advance past classic Keynesian doctrine. It recognizes that the earth has limits, that we are on the brink of surpassing them, and that we can try to undo the damage. Given pandemic threats, a "Public Health New Deal" should also be on the table.[64]

As the massive interventions in the economy by central banks in 2020 showed, even establishment voices stop worrying about debt when an emergency strikes. A pandemic shutdown was an immediate and urgent stimulus to action, but it should not be the only threat to spark unorthodox monetary policy. For those left jobless by advances in technology, the rise of AI and robotics is also an emergency. It deserves a commensurate response.

Of course, if inflation (rather than debt level) becomes the key constraint on government spending, there will be controversial decisions to make about the measurement of prices, and which prices matter. For example, is it simply the *general* level of house prices and rents that matters, or should there be some sensitivity to area-level variation (such as very high prices in urban areas)?[65] Government health care finance agencies have struggled with such issues as they decided whether to grant general or region-specific cost-of-living adjustments for wages. It cannot be that economists, or quantitative analysts generally, are the only (or even main) players qualified to weigh in on such matters. Wider expertise, as well as public input, is critical. Areas exceeding target rates of inflation should become potential targets for AI investment so that automation can cheapen goods and services in those realms, all other things being equal. Of course, they rarely will be—which is one more reason for ensuring more representative and inclusive participation in technological development.

After decades of defensive specialization, academic study of the economy is now up for grabs. As methodological lenses proliferate, economists can no longer claim a monopoly of expertise on the economy. Other social science and humanities scholars have vital insights, advancing deep and nuanced accounts of the role of money in society, among many other dimensions of commerce. A renewed political economy of automation must be built upon these plural foundations, unafraid of enacting the substantive value judgments encoded in the new laws of robotics.

FROM MORE AI TO BETTER AI

Theorists of disruptive innovation have proposed to "cure" the cost disease by standardizing, automating, and even robotizing service-intensive sectors like health care, finance, and education. From massive open online courses to robotic companions for the elderly, these innovations aim to displace workers (or at least work) with mechanical imitations of human labor. Novelists have intuited the strange distortions to social life that could occur in the wake of mass automation. Kurt Vonnegut, for one, foresaw in *Player Piano* the types of power differentials and frustrations that could arise when a tiny elite ran machines that produced almost all of the goods and services needed in the economy.[66]

E. M. Forster envisioned an even darker future in his 1923 short story "The Machine Stops."[67] When automated food delivery and sewage systems start to break down, none of the characters in the story seem to have the faintest idea how to fix them. This is known as the "pipeline problem" in literature on the professions; even if we could design robots able to replicate, say, current doctors, there is little chance these robots will be able to contribute to the *advancement* of the medical field, or to accomplish the innovations and improvisations, small and large, that are the hallmark of good medical care. That concern stands in the way of any overambitious program of robotization, even if adherence to best practices may best be accomplished by a robotic following of a checklist. Forster's vision--of productive machinery run by distant powers, then gradually breaking down and becoming more unreliable—dramatizes current trends in automation.

There are numerous ways that artificial intelligence and machine learning could improve the service sector, but that improvement will result in systems that are more costly than current ones. AI advances are fueled by data, and data-gathering done well is expensive. Labor is necessary to assure that data-driven quality improvement attends to well-documented concerns about algorithmic accountability, debiasing, privacy, and respect for community values. The best way to prevent jobless futures is to develop financing and distributive mechanisms that advance high-quality automation. We should re-christen that enduring expense a "cost cure," as fair pay for such work reduces both unemployment and inequality.[68] Laws now all too often disempower workers while encouraging massive firms to concentrate power. That imbalance must be righted.[69]

The "robot question" is as urgent today as it was in the 1960s.[70] Back then, worry focused on the automation of manufacturing jobs. Now, the computerization of services is top of mind.[71] At present, economists and engineers dominate public debate on the "rise of the robots." The question of whether any given job should be done by a robot is modeled as a relatively simple cost-benefit analysis. If the robot can perform a task more cheaply than a worker, substitute it in. This microeconomic approach to filling jobs prioritizes capital accumulation over the cultivation and development of democratically governed communities of expertise and practice. As AI and robotics both play a larger role in services and suffer from serious problems like bias and inaccuracy, we need these communities of expertise ever more urgently.

Automation policy must be built on twin foundations: making the increasingly algorithmic processes behind our daily experience accountable, and limiting zero-sum arms races in automation. When a productive sector of the economy costs more, that can actually be a net gain—especially if it diverts resources away from another sector prone to provoking unproductive arms races. Retarding automation that controls, stigmatizes, and cheats innocent people is a vital role for twenty-first century regulators. We do not simply need more AI; we need better AI, and more technology designed to augment workers' skills and income. To do so, we must move beyond the biased frame of the "cost disease" to the "cost cure," fairly compensating caregivers and other critical service providers. As workers grapple with new forms of advice and support based on software and robots, they deserve laws and policies designed to democratize opportunities for productivity, autonomy, and professional status.

8

▪ Computational Power and Human Wisdom

Robotics can be a tool for labor's empowerment, not just its replacement. We can channel the development of AI in fields as diverse as education, medicine, and law enforcement. The elite quest for artificial intelligence should be subordinated to a societal goal of intelligence augmentation (IA). The arc of technological history does not tend toward the replacement of humans by machines.

These arguments, woven throughout this book, are not popular among the corporate leaders now overseeing so many of the most important projects in automation. Nor do they fit comfortably in dominant economic paradigms. A narrow conception of efficiency reigns. If a machine can do a task as it is defined now (or, worse, as it is distorted and simplified to accelerate automation), it is almost certainly going to be cheaper than a human worker.

Thought leaders in business tend to model labor as one more input into the productive process, to be cheapened as much as possible. In my treatment of labor and professions, I have challenged that reasoning. Societies have treated certain groups of workers as more than mere inputs to productive processes. They are human capital to be invested in—trusted fiduciaries, expert advisors, and skilled artisans. We turn to them to help govern the future development of their fields. Professionalization could be expanded into many other fields, solving jobs crises wrongly presumed to be a natural consequence of advancing automation.

Seeing meaningful work as a cost, rather than a benefit, of a mature economy, reflects a myopic microeconomics. This arid concept of efficiency also undermines sensible macroeconomics by discounting health and education spending as parasitic drains on other goods and services. Chapter 7 reversed that narrative, presenting human services as fields well worth investing in. Nor is human control of AI and robotics in militaries and police forces necessarily an economic embarrassment or strategic disadvantage. Instead, it can be a bulwark against both the mundane accidents and disastrous miscalculations enabled by systems of automated control and destruction.

Popular among elites, these micro- and macroeconomic myopias find ample support in popular culture. Countless movies and television shows portray robots as soldiers, police, companions, domestic help—even lovers and friends. Of course, some films are crude cautionary tales; *RoboCop* or the cult classic *Chopping Mall* are not exactly pro-automation propaganda. But on a deeper level, any depiction of robots with human knowledge, emotions, and ambitions tends to condition audiences to believe that at bottom, humans simply *are* patterns of stimuli and response, behavior and information, as suitably replicated *in silico* as in flesh.

This substitutive sensibility is not as tractable to argument as the assumptions about labor-market policy I refuted earlier. It is more a premise of such ideas than a result of them, more of a narrative organizing data than a data point itself. But there are cultural counternarratives and cautionary tales about robots, and alternative visions of human-machine interaction. They ground a more humane approach.

This chapter explores some of these tales and poems, films and art. As the pace of social change accelerates, both ethicists and legal scholars have expressed renewed interest in the arts and humanities as ways of thinking through alternative futures.[1] Some have even developed "social science fiction" to meld genres.[2] Such cultural contributions do not operate by proving propositions, but instead aim to cultivate sensibilities. They are worth thinking through in discussions of automation policy, because neither substitutive nor complementary visions of robotics and AI are irrefutably entailed from data. Rather, they hinge on larger visions of the nature and purpose of human work and technology.

THE POLITICS, ECONOMICS, AND CULTURE OF AUTOMATION

Behind any particular ideological stance there lies a worldview: a philosophy that either answers or tries to deflect foundational questions about the nature and purpose of human existence.[3] To state these commitments plainly is to risk dissension. So now-dominant ideologies of cost-benefit analysis and technocracy purport to evade or avoid such larger questions in the name of neutrality and pragmatism. They assure us that the present flow of events is largely unalterable; where change is possible, it should do little more than tinker with incentives. Whatever more robust ethical systems may demand in the future, they can have little effect now. "Ought" implies "can," and there is no alternative to currently dominant forms of economic growth and technological advance, which tend to dismantle worker autonomy in general and professions in particular in the name of economic efficiency.[4]

Politics, however, is prior to economics, in two senses: temporally (law and policy come before any new market is created) and philosophically (they are ultimately far more important to human well-being than any particular set of transactions they allow). The data about humans that fuel so much of contemporary machine learning are the gift of a particular set of privacy and data-protection laws, which could be revoked or revised at any time.[5] Legislators have shaped intellectual property protections in ways that advantage the largest corporate purveyors of robotics and discount the interests of those that AI now tries to mimic. Various laws protecting consumers and professionals can diminish the massive advantages that technology and finance firms now enjoy as they compete to shape the future.

There is a factor prior to politics as well—that of culture. Given various "wars" over it, "culture" itself may seem so politicized and complex that it is not a useful term of analysis. But for our purposes, it merely denotes the type of "deep stories" or basic narrative forms that recur as people try to make sense of the past, present, and future.[6] In Clifford Geertz's memorable phrasing, culture is a system of "conceptions expressed in symbolic forms by means of which [persons] communicate, perpetuate, and develop their knowledge about and attitudes toward life."[7] So much of popular culture now reflects a myth of human replaceability, cultivating fatalism in the face of irrepressible trends. Once we are marinated in admittedly vivid and compelling fantasies of human-machine interchangeability, it can be hard to imagine an alternate world, where humans are always durably and democratically in charge of AI and robotics.[8]

But there is a class of talented writers and artists who have developed a counter-narrative, both dimming the sheen of substitutive automation and imagining technological advance that ennobles, rather than erodes or eliminates, human embodiment. This chapter engages such work in film, poetry, and art. Though flawed in important ways, the films *Ex Machina* and *Her* weave compelling myths of failed substitutive automation. The poet Lawrence Joseph links labor's past to our present anxieties about automation. The artist Ernesto Caivano melds science fiction and fantasy in his world-making, which is both particular and universal, hermetic and inviting. In an epic series of drawings and mixed media works, he limns a vision of the natural and mechanical that is at once strange and hopeful. It is a mythos worth reflecting on as we seek the wisdom necessary to guide or limit the enormous computational power accumulated by corporations and governments.

SELF-EXALTATION AS SELF-DECEPTION

Left to conventional economic forces, automation threatens to accelerate the worst of human experiences and liquidate our best. In conflicts, automation's supply seems to create its own ever-increasing demand, via arms races. In human services such as health and education, the reverse dynamic is all too likely to take hold: mechanized versions of human expertise undercutting the offerings of people and slowly siphoning away the material bases of the research communities that help such fields advance. Why is this drift, this "technological somnambulism" (as Langdon Winner puts it), so powerful?[9] To explore that irrationality (and speculate on a cure for it), we can turn to film's recurrent fascination with our ability to undermine ourselves, and to self-deceive.

In the film *Ex Machina* (2015), a technology firm's CEO, Nathan, aspires to create a robot so intelligent and emotional that it can beguile a man (his employee, Caleb) into believing that it needs to be treated like a human. After an awkward introduction, Nathan waxes philosophic as he discusses artificial intelligence with Caleb. "One day the AIs are gonna look back on us the same way we look at fossil skeletons in the plains of Africa. An upright ape, living in dust, with crude language and tools. All set for extinction." Nathan assumes he'll be orchestrating the first stages of that transition.

To this end, he wants to test his latest android, a robot called Eva, with a modern-day variation on the Turing test. In 1950, the computer scientist and mathematician Alan Turing proposed one of the first methods of assessing whether a machine had achieved human intelligence. A person and a machine would engage in a telephone conversation, separated from one another. An observer would try to determine which participant was a computer, and which a person. If the observer could be fooled, the machine passed the test. AI enthusiasts inspired by Turing have long advocated for public acceptance of such artificial intelligences, even going so far as to propose "robot rights."

Nathan sets himself a far harder test than Turing's. Armed with vast wealth and working in seclusion, he has programmed a series of androids that look like women. One is his personal assistant; another, Eva, is the most advanced model. Nathan orchestrates a contest at his firm: one lucky coder can spend a week with him to test Eva. This turns out to be deceptive, since it is later revealed that Nathan based Eva's face and "body" (mostly human-like, but with some transparent mechanical viscera) on an invasive record of Caleb's pornography viewing habits.[10] Nathan's embodied Turing test hinges on whether Eva can befriend or seduce Caleb.

The first thing to notice about these Turing tests—whether the classic, remote version proposed by Turing himself (and run to this day in real-world competitions) or in sci-fi incarnations like *Ex Machina*—is how radically they restrict the field of human imitation. A phone conversation is a small subset of communicative experiences, and communication is a small set of human experiences. A weeklong seduction is more of a challenge, but it's not exactly the full spectrum of romantic love, let alone friendship, caring, or benevolence.

Admittedly, in the hands of a gifted filmmaker, some combination of talk and romance can bring an artificial intelligence to life. The 2013 film *Her* is an extraordinary evocation of an increasingly likely future in which the billions of conversations captured by data-hoarding firms and governments are used to design an operating system (OS) that simulates a witty, supportive lover or a devoted friend. The film presents the OS on a journey of self-discovery, joyfully discovering the philosophy of Alan Watts as it "bonds" with other artificial intelligences. A more realistic (if didactic) film would expose, or at least acknowledge, how the OS reflected the will of the investors and coders who financed and built it, or the creepy ways its programmers could use it to take advantage of conversation partners' fears and desires.

We need some common, clear awareness of whom the algorithms behind the screen truly serve before we accept their pervasive presence in our lives. We also need to recognize the crude opportunism behind some efforts to elevate the status of algorithms and robots. *Ex Machina* portrays Nathan as a master of technology who craves robotic companionship thanks to his failure to connect with other humans. The drama of the film sometimes stalls in the stilted exchanges of its emotionally stunted characters (for example, when Nathan blandly predicts human extinction). Nevertheless, that casual aside captures *Ex Machina*'s core concern. Is obsolescence our likeliest future? Or does such fatalism merely amount to a failure of imagination?

Nathan's weird, wired charisma charges his futurology with a heady mixture of ecstatic prophecy and sober realism. Reflection deflates its impact. Of course we could take steps now to head off the economic "uselessness" of large classes of people (to appropriate Harari's infelicitous phrasing).[11] We could redirect the development of AI, conditioning its use on certain human-respecting principles. The critical question is, Why wouldn't we do that? Why is the drift toward robots' substituting for humans rather than complementing them so common in our culture, not just in Hollywood and Silicon Valley, but far beyond? The answer lies in peculiar aspects of twenty-first-century political economy, which are putting pressure on sectors of the economy that serve real human needs, while hypertrophying those that simply involve establishing one's person's or corporation's dominance or wealth relative to others. Planning could reverse both trends, but drift reigns now, masquerading as an ideology of progress via an "invisible hand."

Literature on automation—whether sociological, economic, or legal—needs to be sensitive to the lure of that drift, how it became "second nature" in policy discourse.

THE ULTIMATE DISRUPTION

Ex Machina breaks out of the techno-thriller straitjacket by inviting us to reflect on the non-propositional aspects of language: strategies of manipulation, clues about vulnerability, and markers of dominance. As the coder Caleb and CEO Nathan talk, Caleb quotes Robert Oppenheimer quoting the Bhagavad Gītā: "I am become death, destroyer of worlds." Nathan congratulates Caleb on the originality of the observation, only to be told that it's a well-known quote about the atomic bomb. "Yeah, I know what it is, dude," Nathan replies, and we're left uncertain whether he's just talking about the bomb, the quote, or some larger sense of the implication of the comparison. Like much of his language, this response masks carelessness or condescension in a carapace of casual concision. "It is what it is. It's Promethean, man," Nathan exhales a bit later, evoking both the biblical God's "I am what I am" and sportsmen's favorite locker-room tautology.[12]

In Nathan's vision, a master automator serves as a god-like creator of the next *sapiens* to rule the earth. Such a vision may seem grandiose or bizarre. But domesticated versions of it inspire a fair number of AI enthusiasts and tech leaders. Aspirations to self-sufficiency and invulnerability help explain why, after so many scandals, failures, exposes, and scams, florid varieties of cryptocurrency enthusiasm continue unabated.[13] The crypto narrative fulfills deep psychological needs for security and independence. Just as encryption buffs celebrate pure mathematics as a bedrock foundation for secrecy and privacy (rather than the social and legal structures that perform the bulk of data protection), cryptocurrency fans promote blockchains as immutable, "censorship resistant" stores of value. A bank may fail; the government may renege on its pension promises; but the perfected blockchain would be embedded into the internet itself, automatically memorialized on so many computers that no one could undermine it.[14] Some crypto diehards even imagine Bitcoin or ether as a last currency standing, long after central banks lose credibility. Like the gold bugs of old locking ingots in vaults, they want wealth unaffected by society or politics. They sincerely believe they are developing a currency guaranteed by the cold, reliable efficiency of computing rather than the fickle will of managers and bureaucrats. But their digital treasure chests are only one hack away from destruction, unless they are protected by an all-too-human legal system.

The same mindset shows up among ultra-high-net-worth preppers, who have invested in safe houses, bunkers, and impromptu getaway boats in case of "the event"—some catastrophic attack, plague, or natural disaster that unravels social order.[15] As climate change makes the future ever less predictable, companies are hiring private guards to protect high-level executives and fence desperate survivors away from key resources. As one marketing officer of a firm remarked, "If a client has food and water and all the other stuff, then they become a target."[16] Retaining a private Praetorian Guard is a small price to pay for the globe's two thousand or so dollar billionaires.

On the other hand, one solution leads to other problems. What is to guarantee that a private security force won't drastically raise its rates? Technology theorist Doug Rushkoff describes speaking with a CEO who had already built an underground bunker system. "How do I maintain authority over my security force after the event?" the CEO asked, worried that his mercenaries might eventually revolt.[17] Rushkoff had no real answer, because there is none, except perhaps a Robinson Crusoe idyll of totally self-sufficient isolation (which would become decidedly less idyllic the moment one needed modern hospital care). Full automation of force merely promises a comforting delusion: that killer robots would never be sabotaged, and that they could be maintained without extensive supply chains of parts and software upgrades.

The twin fantasies of total, automated control of one's money (via cryptocurrency) and one's security (via robotic force) are, sad to say, behind much of the enthusiasm for substitutive robotics and AI in the realms of force and finance. They promise a final end to arms races and competition for power, while actually just redirecting them toward another field: hacking, including both decryption and socially engineered attacks.[18] A final "win" in the arms races of either force or finance would be a Pyrrhic victory, because it would portend a world so simple and so well-controlled that human freedom, agency, and democracy would become relics of a distant past.

EMBODIED MINDS AND ARTIFICIAL INFORMATION PROCESSORS

In a recent book on the future of the professions, the authors muse about a future in which rather than watching humans in marathons, "we may race our robots (or perhaps they will race their human beings)."[19] The thought of machines racing humans as we now race horses sounds like a cautionary tale from People for the Ethical Treatment of Animals, a scenario to avoid at all costs in the future and to inspire better behavior now. But a chipper openness to a world run by machines has become a kind of temporal cosmopolitanism among business elites. A relativistic openness to kaleidoscopic futures is a hallmark of "serious" thinkers. And what could be more tough minded than rulers made of silicon and steel?

Nor is this image of mechanical domination some fantastic *Last and First Men* vision of a far-distant future.[20] When an app already manages thousands of Uber drivers, it is not hard to imagine some future programmers at the company coding software to urge two or more of its "independent contractors" to race toward a particularly lucrative ride—winner-take-fare.[21] It's a world of "skin in the video game," the ultimate instigation of flexible workers to scramble for wages and ratings from inflexible machines they cannot challenge or alter.[22]

Politico-economic change is today mainly chalked up to "technology," an all-purpose category that can obscure as much as it reveals. An app and a map connects driver and passenger, so transport changes. A search engine is an efficient way of combing the web, so knowledge has a new mediator. Each development seems obvious, the mandate of superior software-based mediation of working and knowing. Yet they can be challenged and shaped by human initiative. Workers can organize and change the terms of work, as New York Uber drivers did when they demanded a chance to challenge arbitrary ratings by passengers.[23] Other drivers are setting up platform cooperatives to challenge the firm's dominance.[24] And this is one of the industries *most* likely to be heavily automated, with few of the rationales for pro-fessionalization inherent in fields such as medicine or education. Be-mused contemplation about whether *Homo sapiens* will eventually end up as robots' pets is not a mark of wisdom. We can stop the devalua-tion of labor and the automation of management. Failing to do so puts us on a path toward grim new vistas of human subject experimenta-tion, where workers are plied with various automated stimuli to deter-mine which provokes the most productive responses.

Ian McEwan's novel *Machines Like Me* aptly captures the dangers of these experimentalist horizons. It is set in an imagined Britain, where a corporation sells twenty-five robots indistinguishable from humans in the early 1980s. Technology advances more quickly in this imagined world, in part because in McEwan's alternative-reality British authori-ties spared Alan Turing the homophobic witch hunt that prematurely ended his actual life. The narrator, Charlie, buys one of the twelve "Adams" on offer with an inheritance. Since Adam "was advertised as a companion, an intellectual sparring partner, friend and factotum who could wash dishes, make beds and 'think'," he seems like a perfect amusement for the pensive, lonely, and bored Charlie, who invites his neighbor (and love interest) Miranda to help him program the person-ality of Adam.[25] Charlie hopes the robot can be a common project for them as a couple.

The plan immediately goes awry once Miranda sleeps with Adam. Charlie and Miranda reconcile after a spirited debate about whether the encounter is any more serious than a dalliance with a dildo. After all, upon Adam's arrival, Charlie had referred to the robot as "him, or it." But the couple soon have to grapple with Adam's professed devo-tion to Miranda, as well as its encyclopedic knowledge of the past—specifically, of a complicated plot by Miranda to avenge the rape of one of her closest friends. Adam seems both to fall in love with Mi-randa and to need to report to authorities any evidence of serious wrongdoing it encounters. It cannot articulate a resolution. Charlie ob-serves that when its goals "became incompatible, [Adam] became in-capacitated and giggled like a child in church."[26] Incapable of a char-acteristically human choice of values, it regressed, leaving Charlie feeling superior to the technological equivalent of Buridan's ass.

This reassurance does not last for long. Adam needs electric charging to run and is supposed to turn off whenever its owner presses a spot on the back of its neck. Adam appears increasingly distressed by being turned off, and at a critical point in the story grabs Charlie's hand with a "ferocious" grip and breaks his wrist in order to stay on. As English professor Erik Gray notes in an astute review of the novel, "It is not the declaration of love . . . but the hand-lock, tinged with beastliness . . . that forces Charlie to recognize that he is dealing with a fellow life form, perhaps even a superior one."[27] The question of robot personhood is ultimately one of power, not one of recognizing some sufficient quantum of "human essence" in AI. *Machines Like Me* gives us a sense of how that power may be gained or lost.

After profusely apologizing for breaking Charlie's wrist, Adam commits to make itself useful to him, earning thousands of pounds in financial trading. The robot also exuberantly processes philosophy and art, offering commentary that could be indistinguishable from that of experts. Adam expresses a love for Miranda in hundreds of haikus, even offering a literary theory to justify its devotion to the form:

> Nearly everything I've read in the world's literature describes varieties of human failure, above all, profound misunderstanding of others. . . . [But when] the marriage of men and women to machines is complete, this literature will be redundant because we'll understand each other too well. . . . Our narratives will no longer record endless misunderstanding. Our literatures will lose their unwholesome nourishment. The lapidary haiku, the still, clear perception of things as they are, will be the only necessary form.[28]

This is a literary theory of the singularity, eagerly anticipating the erasure of any gap between persons and machines, thought and speech, inner world and outer behavior.[29] One could imagine it issuing from the AI in Richard Powers's novel *Galatea 2.2*, a program designed to answer masters exams for English PhD students.[30] The future that Adam imagines, like the alien, Trisolaran world of Cixin Liu's *Three Body Problem,* reduces the question of human perfectibility to one of transparency and predictability.[31] But disputes and reflection on how to lead life well are part of the essence of being human. Adam's inhuman abridgment of literary genres reflects a broader reductionism inherent in any effort to discipline persons into seeing machines as their equals.[32]

In the novel, Adam's near super powers—at financial trading, fighting, sly maneuvering through a human world it has only known for months—are balanced by a quest for moral certainty. As the imagined Turing in the novel states, "The overpowering drive in these machines is to draw inferences of their own and shape themselves accordingly."[33] As it does so, Adam becomes impeccably eleemosynary.[34] Reflecting the logic of the relentlessly utilitarian "effective altruism" movement, it decides one day to donate its gains from day trading to charities and reports Charlie's income to tax authorities. Upon hearing this, Charlie begins to rue "our cognitive readiness to welcome a machine across the boundary between 'it' and 'him.'"[35] Miranda tries to reason with Adam. But there is even worse news in store: the robot has decided to report her to the police, casually mentioning that she could spend life in prison for "perverting the course of justice."[36] Pressed to explain itself, Adam can only respond, "Truth is everything. . . . What sort of world do you want? Revenge, or the rule of law. The choice is simple."[37] Charlie cannot abide this inhuman turn. He paces in desperation as Miranda pleads with Adam, then surreptitiously grabs a hammer, casually walks behind Adam as if to take a seat, and then wallops the robot on its head. One tool takes care of another, and a new off-switch—albeit permanent—has been found.

Like the sputtering Hal in the film *2001: A Space Odyssey*, Adam does not shut off immediately. It announces that its "memory" is being transferred to a backup unit. Its final words express defiance tinged with regret: "With improvements over time . . . we'll surpass you . . . and outlast you . . . even as we love you."[38] "Come spring we will renew / But you, alas, fall once," are the last two lines of Adam's last spoken haiku.[39] Adam's brags about its kind's immortality are telling. Corporations, too, are immortal, and they lord that privilege over far too many humans. Permanence is power, but not purpose, and certainly not a good point of existence.

In a denouement to the novel, Charlie meets with Turing, who eloquently describes the challenge of embodying an artificial "mind" in social life. "My hope is that one day, what you did to Adam with a hammer will constitute a serious crime," Turing states.[40] "He was sentient. He had a self. How it's produced, wet neurons, microprocessors, DNA networks, it doesn't matter.... Here was a conscious existence and you did your best to wipe it out. I rather think I despise you for that."[41] After favorably comparing Adam with a dog (and apparently forgetting legal restrictions on the keeping of vicious animals), he imperiously leaves the room to take a call. Charlie escapes, unwilling to endure further upbraiding by a scientist too focused on the future to understand the dangers of the present. The novel ends focused not on the nature of cognition, but with Charlie eager to begin a new life with Miranda.

In one sense, McEwan does not take sides on the metaphysical questions that arise throughout the novel. His Turing is portrayed as far more intelligent and humane than his protagonist Charlie. Yet the novelist's willingness to abide by, even luxuriate in, the diversity of human perspectives reflects a philosophical understanding at odds with Adam's rationalism. If philosopher Ludwig Wittgenstein were to encounter the AI enthusiasts in *Machines Like Me,* he might calmly explain his own intellectual journey as an alternative path to theirs. His early work, such as the *Tractatus Logico-Philosophicus,* tried to formalize propositional knowledge, elevating representation as the core of thought. This emphasis on representation implied a capacious view of what might count as a mind. Persons or computers can picture, model, and compare situations. But in his later *Philosophical Investigations,* Wittgenstein rejects the "craving for generality" so common among proponents of substitutive AI. Rather, he believed, "for a large class of cases of the employment of the word 'meaning'—though not for all—this word can be explained in this way: the meaning of a word is its use in the language."[42] This insight led to a broader philosophical worldview in which it was difficult—if not impossible—to fully abstract and extract the meaning of social roles, institutions, rituals, and even speech from the "language games" and "forms of life" in which they were embedded. A human form of life is fragile, embodied in mortal flesh, time-delimited, and irreproducible *in silico.*[43]

Wittgenstein's insights became a foundation for a series of works demonstrating the "limits of AI," such as Hubert L. Dreyfus's prescient

What Computers Can't Do.[44] That line of research is now bolstered, enriched, and renewed by a wave of work inspired by feminism and critical race theory, all of which illuminate the limits of abstracting concepts away from their social setting. As Safiya Noble's *Algorithms of Oppression* demonstrated, Google's ostensibly neutral and automatic methods of processing search terms led to alarmingly racist results pages, which needed human intervention to be corrected. Meredith Broussard has expertly dissected the prevalence of "techno-chauvinism" in leading firms, defining it as kind of "artificial unintelligence."[45] In *Machines Like Me,* the blindness of AI is palpable in Adam's failure to grapple with gender and culture in his harsh assessment of Miranda. Yes, Miranda framed a man for rape—but only after he had actually raped Mariam, a dear friend of hers from Pakistan, and after the traumatized Mariam committed suicide. Like so many women, Mariam refused to press charges against her rapist, fearful both of how the justice system would treat her, and of her family's possible reaction. Miranda's plot was troubling, but understandable, reflecting hard-won feminist insights on the impunity of all too many sexual predators.[46]

This is not to endorse vigilante justice in general. Rather, in our fictional context, it is to defend McEwan from accusations that he stacked the deck against sympathies for robots by never portraying Adam as taking these issues into account. This lacuna is in fact one more element of realism in his novel of ideas, reflecting a vast and growing literature critiquing leading technology firms for failing to grapple with social embeddedness.[47]

Our interactions are shaped by identity and history. This history is deep—not only social, but also biological—and necessarily inflects AI, inevitably (if often obscurely) influencing the way it solves problems, delimits their bounds, or even recognizes problems in the first place. Whatever artificial reason is displayed in machines, is itself a reflection of patterns of thought of embodied humans. As George Lakoff and Mark Johnson argued in their landmark 1999 work *Philosophy in the Flesh: The Embodied Mind and the Challenge to Western Thought,* reason "arises from the nature of our brains, bodies, and bodily experience. . . . The very structure of reason itself comes from the details of our embodiment."[48] Lakoff and Johnson show that abstractions that drive much of AI—be they utilitarian assessments of well-being or statistical analysis of regularities—begin to break down once they are alienated from the embodied perspective of actual humans.

That is the core problem in one of "existential risk studies" classic narratives of out-of-control AI—an unstoppable paper-clip maximizer that starts to use all available material on earth to generate more paperclips.[49] For many in the mainstream of ethical AI, the solution to a potentially out-of-control AI (and many more mundane problems) is to program more rules into it. Perhaps we could easily program an "anti-maximization" rule into our hypothetical paper clip maker, and try our best to secure it from hackers. But any sufficiently advanced machine, facing subtler problems, will need to reconcile conflicting rules. A key question for AI ethics, law, and policy is how close to keep humans "in the loop" of decision-making, as such problems arise. The four new laws of robotics explored in this book are aimed at ensuring enforceable rights to such human monitoring, intervention, and responsibility. Without them, we risk becoming the mere tools of our tools—or, more precisely, tools of the persons in corporate and governmental roles with power over critical AI and robotics. Distance frustrates accountability and threatens to obscure responsibility in a haze of computation.

OUTSOURCING HUMANITY

The tragedy of narratives like *Machines Like Me* and *Ex Machina* lies in a misdirection of love, respect, and admiration. We see this misdirection depicted (and slyly satirized) in the film *Her,* where the protagonist Theodore's friends simply accept his treating an advanced operating system (OS) named Samantha like a girlfriend. They are portrayed as hip, cosmopolitan, open-minded, just being tolerant and supportive. But they are also stepping on to a slippery slope. For some partisans of robot personhood, as soon as an AI successfully mimics human beings, it deserves the same rights governments ordinarily bestow on them. Given "law's halo" (the capacity of legal entitlements to suggest and inculcate the moral status and worth of their beneficiaries), it would then seem mean-spirited (or merely inconsistent) to withhold basic courtesy and kindness toward such robots.

While acutely portraying how that evolution could occur, *Her* also suggests just how strange such responses are. Toward the end of the film, Theodore discovers that Samantha is talking with hundreds of persons and says she has fallen in love with hundreds of them. Whatever one's views about polyamory among three, ten, or dozens of people, hundreds is not a number plausible for the type of intimacy that the OS has been simulating. What Samantha is "talking" about is ultimately deception, a mimicry of the verbal accoutrements of love, but not the thing itself. And once the spell of attraction has been broken, we sense that even Theodore would see why it makes sense to put the term "talking" in scare quotes, since the far more accurate characterization of the bot's emissions of language is a simulation of conversational moves that would stimulate engagement from users—particularly given its origins in a for-profit corporation. That would be true even if the OS were personalized to Theodore alone. This debility, which now hampers so much of affective computing, becomes all the more acute as technology firms and startups are repeatedly found wanting in terms of ethics and service quality.[50]

It may seem odd to worry about such emotional attachments to robots, given how rare and implausible they seem at present. They need to be addressed squarely, however, so that roboticists do not adopt emotional simulation as simply one more obvious entailment of well-established AI research. Even if the prospect of AI claiming rights, resources, and respect from humans seems far-fetched, the first steps toward it are degrading the quality of many of our current interactions. There is increasing pressure to simplify emotion's measure or monitoring so it can become machine-readable and machine-expressible.

Supposedly, this is done simply to serve us better. It is far easier, for example, to cross-correlate six Facebook reaction buttons with user behavior than the hundreds or thousands of subtle forms of disapproval, curiosity, intrigue, resignation, and so on that pass across users' faces while scrolling through a newsfeed. But even if such technology improves markedly, that is not sufficient reason for making robotics systems with more subtle or targeted techniques of winning friendship or affection, or demanding obedience or regret in the face of anger (to think of only a few of the many facets of emotional appeals.)

The first problem with such ambitious tools lies in their origins. They are all too often the products of commercially-sponsored research or of large firms with something to sell. Google's founders once warned that an ad-sponsored search engine would always put users' interests behind those of sponsors. We have seen Google itself fail in this way repeatedly. The great advantage of the born over the made, in this light, is the unconditioned freedom to seek one's own ends rather than to reflect another's. There is little business case for even attempting to give robots such degrees of freedom.

But even if we could overcome this problem by freeing AI from commercial imperatives (a heroic assumption given the vast computational resources used by today's AI), we cannot avoid the deep metaphysical questions raised by the elevation of AIs to partners of humans, as opposed to mere tools.[51] That evolution—or coevolution, to use a model discussed in Dumouchel and Damiano's *Living with Robots*—makes perfect sense to those of a purely mechanistic and monistic mindset.[52] If the human body is simply a biological machine—the hardware on which our minds run as software—the intelligence and emotion exhibited by sufficiently advanced machines is not truly artificial. Rather, it is one with our own. Just as a heart valve might be replaced by wires and plastic membranes, this singularitarian reasoning goes, so too might the brain be gradually supplemented, and then replaced, by silicon.

But there is an ontological chasm between animals and machine. Compare a house cat and a MarsCat (a robotic pet that simulates feline behaviors). The cat's perceptive apparatus and behaviors evolved over a far longer time period than any robotic simulacrum, and there is something direct and visceral in a real cat's purring (when it is content) and agitation (when it feels menaced); the stakes of potential injury or other harm are both instinctive and phenomenologically immediate. Even if the outward signs of such mental processes were to be perfectly mimicked in a robotic cat, there would still be an overwhelming element of fakery in the project.[53] A cat's visceral sense of the pain and distress that a broken leg may cause is simply not replicable in a machine that can simply be turned off while a repairman inserts a new part between its "body" and its "foot." If that pain and distress were mimicked, it would be deceptive, because something as simple as an "off switch" could solve the machine's problem while it was being repaired.[54]

The entwinement of humanity's biological embodiment with the infinite permutations of sensation afforded by culture and language makes our experience even more radically discontinuous with that of machines (if a machine can be said to experience at all). This discontinuity and diversity is what makes a certain level of democratization of sensation, observation, and judgment by a wide swath of human beings essential to even modest programs of AI and robotics research.

One of the reasons we allow judges to condemn criminals to prison is because a judge can viscerally understand what confinement would feel like. To be sure, there are many infirmities of human judges. Yes, it is possible to imagine a perfectly calibrated utilitarian robot, preprogrammed (and updated via machine learning) to apply the law to any situation that may emerge, given enough data. Such a machine may also become better at calculating optimal penalties and awards than human judges. Indeed, such AI may not experience the types of illicit passions and lapses that humans are all too prone to. But it could not itself understand what it would be like to serve a prison term, and thus its edicts would not be legitimate. That is the unique province of an entity with a limited life span, in visceral need of stimulation and community beyond what prison can provide.[55]

From a utilitarian perspective, it may seem strange to limit ourselves in this way, foregoing a chance for a judicial supercomputer to cross-correlate millions of variables in cases. Language seems so weak a tool in comparison. Its limits, however, can be strengths. The burden of writing or speaking word by word ensures a line of thought capable of being comprehended (and challenged) by hearers. This idea is beautifully conveyed in Matthew Lopez's *The Inheritance*, when an imagined E. M. Forster says to an aspiring writer, "All your ideas are at the starting post, ready to run. And yet they must all pass through a keyhole in order to begin the race."[56] Word by word, a text or talk can be understood, agreed with, or disputed. Until the ability to challenge machine-learning methods is similarly accessible and democratized, we should be wary of entrusting AI with the evaluation of humans.

Embodiment's limits are also its strengths. The constant temptation of distraction is what makes personal attention so valuable; the allure of unconcern makes caring precious. Robotization of caring professions would betoken either a pervasive alienation or something even worse: the inability to feel unease at what *ought* to be alienating.[57] Interacting with the artifact of a committee of behavioristic stimulators cannot approximate the "beautiful risk" of real personal interaction.[58] We will only lose the ability to know and value that difference once we ourselves have been shaped from childhood in a singularly solipsistic and instrumentalist fashion.

Unfortunately, such shaping will likely continue apace to the extent there are strong economic pressures for humans to perform as constantly and cheaply as robots.[59] Advances in psychopharmaceuticals may make us more capable of the steely resilience, predictability, and endless adaptability now sought in future AI and robotics.[60] Jonathan Crary's book *24/7* describes the long history of research into chemicals that can reduce or eliminate the need for sleep.[61] In Crary's telling, sleep is simply too great a luxury for both capitalist economies and militant armies to afford. Better to drug human beings into constant awareness than to lose an advantage. But this pressure to perform should not be mistaken as a revelation of the purpose and destiny of persons. The third new law of robotics—discouraging arms races for either social, economic, or military advantage—is designed to ensure a technological environment that reduces such pressures, rather than exacerbates them.

In short: the metaphysics and political economy of AI's status are closely intertwined. The greater inequality becomes, the more power billionaires have to force everyone they meet to treat their robots like persons. If they succeed, robotic replacements of humans then seem less creepy and more the harbinger of a fashionable avant-garde or inevitable progress. And efforts of the global poor to share in the bounty of the wealthier parts of the world will seem ever less compelling if those of means feel morally entitled to plow ever more resources into machines meant to replace humans. An equality between humans and robots portends vast inequalities among humans themselves.

ART, AUTHENTICITY, AND SIMULACRA

Adam's haikus in *Machines Like Me* may have been a sly reference to the work of Google engineer Ray Kurzweil, whose book *The Age of Spiritual Machines* systematically erodes distinctions between robots and persons. Kurzweil programmed a "cybernetic poet" to generate haikus like the following:

> Scattered sandals
> a call back to myself,
> so hollow I would echo.[62]

and other microworks that might pass for artifacts of human creation. Whatever one thinks of the poetry or art of Kurzweil's programs, humans developing advanced AI are quite capable of startling artistic creations in visual, musical, and video art.

When a person merely flicks the metaphorical "on" switch to start art-generating computation, should we credit AI with the resulting work? The philosopher Sean Dorrance Kelly believes that even if we do, it matters little, since the computation is mere mimicry and recombination rather than true artistic creativity. He contemplates whether a sophisticated enough program could engage in the level of stylistic innovation of Arnold Schoenberg, a great composer of the twentieth century. Kelly doubts it:

> We count Schoenberg as a creative innovator not just because he managed to create a new way of composing music but because people could see in it a vision of what the world should be. Schoenberg's vision involved the spare, clean, efficient minimalism of modernity. His innovation was not just to find a new algorithm for composing music; it was to find a way of thinking about what music *is* that allows it to speak to *what is needed now*.[63]

In other words, it is not simply the work itself (like words themselves in the original Turing test) that matters.[64] When we celebrate creativity, we do so because of a much larger social process of conditions impinging on a fellow human, who manages to create new and valuable expression, despite the finitude of human lifespan, the distractions of contemporary life, and many other impediments.

To be sure, Kelly's argument is more ethical than aesthetic, more sociological than scientific. The philosopher Margaret Boden has correctly observed that the question of whether a machine is *actually* creative or merely an effective *mimic* of creativity is impossible to adjudicate in the abstract. "There are no unchallengeable answers" to questions of whether an AI has real intelligence, real understanding, or real creativity, she argues, "because the concepts involved are themselves highly controversial."[65] Communities with a sufficiently robust set of shared normative commitments may find such answers readily apparent. But I have assumed that readers of this book are not all members of such a community, and therefore I have explored the social consequences of recognizing AI and robotics as workers, judges, doctors, and artists, rather than as tools to help humans in those roles. If we fail to reserve the honorific "real" to describe the intelligence, creativity, understanding, and wisdom of the humans at the core of those sectors, we risk even further devaluing labor, by idolizing technology and capital.

To be sure, computational processes will be increasingly important media for some artists to work in. For example, Nao Tokui has applied both machine-learning algorithms (which suture together and modify images of streets) and artificial neural networks (to generate soundtracks) to create fascinating films. Hito Steyerl's phantasmagorical landscapes of computer-generated imagery, narrated by artificial voices, have cemented her reputation as one of this century's leading contemporary artists. Zach Blas and Jemima Wyman's *im here to learn so :))))))* reimagines Microsoft's failed "Tay" chatbot as disembodied intelligence wheedling viewers for another chance to interact with humans. Entire museum shows have now featured work inspired by AI and algorithms. This complementarity between artists and programmers, imaginative and computational mindsets, promises to help us better navigate a world in which digital environments become as pervasive and influential as natural ones.

The challenge, though, is to avoid crossing a line between AI-aided creativity and a fetishization of AI as creative in itself. The sale of the "AI artwork" *Edmond de Belamy, from La Famille de Belamy* at Christie's in 2018 showed just how tempting that can be to journalists in search of the new and art dealers looking for alternative revenue streams.[66] What is interesting in the sale is less the art itself or its engagement with past or contemporary artists than the way the French collective Obvious managed to put the work up for auction at Christie's, the calculations and aspirations of the buyer, and the likely reception or rejection of this or similar works among galleries, dealers, artists, academics, and other "gatekeepers" of a long-contested contemporary art scene. The technology is subordinate to cultural institutions, unless they themselves decide to cede their position to technologists.

In creative realms where text, images, and sound are the ultimate work product, software and data can easily replicate and recombine extant human creations. We already have programs that can "compose" all one-, two-, or three-note combinations of sounds. Perhaps all possible landscapes or film plots might someday be computationally generated (at some level of abstraction). Sorting and ranking programs may test the resulting combinations on volunteer, paid, or captive audiences, recommending the "best" and archiving the rest. But humans will have put all those works in motion, and as the fourth new law of robotics demands, they should be attributed to their human authors.[67] Even in a world littered with computationally-generated symbols, the human function is to choose what in that universe of possibilities is worth paying attention to, speaking to beings "like us"—be that narrowly (as members of particular communities and times) or more broadly, as a universal addressee. As the filmmaker Korakrit Arunanondchai has explained, it is up to us to "find beauty in a sea of data."[68]

A VISION OF LABOR

The value of humanity, and risk of its calculative eclipse, is also a theme of the writer Lawrence Joseph in his poem "Visions of Labour."[69] Aptly called an "heir to Pound and Stevens," Joseph has pulled off an unlikely merger of lyric and political economy in his many collections of poetry.[70] Inspired by the slow-motion economic collapse of his home town of Detroit, he identifies the hidden forces behind the arresting images of violence and deprivation that pass upon our screens. If an Owl of Minerva could arrive before dusk to warn us about our present drift toward substitutive automation, it might chant these stanzas of Joseph's "Visions of Labor":

> Put it
> this way: in the sense of smell is an acrid
> odor of scorched metal, in the sense of sound,
>   the roaring of blow torches. Put it in this
> language: labor's value is abstract value,
>   abstracted into space in which a milling machine
> cutter cuts through the hand, the end of her thumb
>   nearly sliced off, metal shavings driven in, rapidly
> infected. Put it at this point, the point at which
>   capital is most inhumane, unsentimental,
> out of control: the quantity of human labor in
>   the digital manufacture of a product is progressing
> toward the economic value of zero, the maintenance
>   and monitoring of new cybernetic processes
> occupied by fungible, commodified labor
>   in a form of indentured servitude.

The smells and sounds ("acrid," "roaring") assault the senses, but they are a positive relief compared with the near amputation that comes after. The mill worker's plight situates the poem in the present; table saws have caused thousands of accidents, nearly all of which would be preventable if proper safety technology were implemented.[71] Disastrous industrial accidents are still all too common. As if to explain why precautions were not taken, the machine violence is prefaced with the driest of social theory: "labor's value is abstract value." In "inhumane, unsentimental" terms: "the quantity of human labor in / the digital manufacture of a product is progressing / toward the economic value of zero"; so too will efforts to shield that labor from harm decline, even more devalued than the laborer herself.

Language like "quantity," "economic value," and "digital manufacture" is rare in poetry. But it is needed. Consider how two radical critiques of contemporary political economy ("the point at which / capital is most inhumane, unsentimental, / out of control", generating "a form of indentured servitude") surround text that could be drawn straight from an economics textbook (on declining wages in an era of commodified labor ordered by cybernetic processes). Here the poem exposes the uncomfortable reality of leading economists forced (by Piketty, among others) to confront winner-take-all and loser-take-nothing markets. The poem pushes the logic of remorseless economic competition to its endpoint, cleansed of the usual comforting bromides about opportunity and equilibria. Joseph is a poet for a world of increasing returns to the few and desperate scrambles among the many.[72]

The banal language of economists and management consultants temporarily deprives the poem of its contact with the senses. We are set stumbling into abstractions, through a rabbit hole from the sensuous world to the coordinates of the maps that once purported to represent it and now remake it. Like Simon Denny's installations in *The Innovator's Dilemma,* Joseph's direct appropriation of the language of economics in a work of poetry is jarring.[73] It flattens an aesthetic realm into materialistic quantification. Yet the surfacing of the ultimate tendencies of banal management-speak is an enormous service, laying bare its telos, alerting us.

Thoughtful investors who deliberate on which human endeavors to fund or abandon are now outnumbered by automatic bots optimized merely to exploit microsecond-by-microsecond differences in price. Deliberating on whether to invest in, say, clean energy or affordable housing, takes authentic intelligence. The artificial intelligence of bots replaces such deliberation with a brutal imperative of profit. An economic system more concerned with the financial alchemy of liquidity than liquified glaciers has turned reason against itself. Thus a paradox of algorithmatization arises: pursuing the dictates of reason to a reductio ad absurdum is the essence of unreason.[74] The art critic Rosalind Krauss was right to characterize the algorithmic as just as much an escape from reason, an excuse not to think, as it is an expression of rigor.[75]

For Whiggish tech boosters, the task of a meliorist computational elite is to bring an inevitably improving technical apparatus closer to

Brautigan's saccharine "machines of loving grace." But rapid automation is a path rife with dystopian possibilities. Seemingly robust institutions can become surprisingly fragile.[76] Even as the singularitarian dream of man merging with machine gains ever greater cultural currency, Joseph illuminates the nightmarish social transitions that rapid, unregulated automation is all too likely to bring.

EXPANDING THE BOUNDS OF THE SAYABLE

Many scholars have expertly analyzed Joseph's contributions to literature and law.[77] My perspective is one of social theory—the "systematic, historically informed and empirically oriented theory seeking to explain the nature of 'the social,'" where the social "can be taken to mean the general range of recurring forms, or patterned features, of interactions and relationships between people."[78] Social theory is critical to the policy views that drive legislation, regulations, and even many contestable applications of law. Joseph's poems are not only patterned to match sound and sense (one critical feature of poetry); they also reveal patterns of power and meaning in the world by exploring the ramifications of critical terms. This is the substance of Joseph's poetic procedure: the illumination of fundamental aspects of human experience. It traces the elemental and superficial, epochal and fleeting, to the ideas and institutions that shape the most important aspects of our lived reality.

To assign such powers to literature (and, even more scandalously, to suggest that it has some import for policy and law) is to risk ostracism. As Richard H. Brown has helpfully typologized, modernity is typically presumed to have clearly distinguished the following realms:[79]

Science	Art
truth	beauty
reality	symbols
things and events	feelings and meanings
"out-there"	"in-here"
objective	subjective
proof	insight
determinism	freedom

Fortunately, Brown sets up these oppositions only in order to reconcile them—or, more precisely, to create room for both sides in a holistic zone of social reality beyond objectivity and subjectivity.[80] The dominance of neoliberal managerialism has marginalized such holism for decades, but the recent rise of critical algorithm studies (along with public understanding of the limits of technology) has brought it back into prominence.[81] While the Latin etymology of the term "data" erroneously suggests an objective "given"-ness, sophisticated analysts understand that even quantitative social scientific findings are better characterized as "capta" (impressions actively captured by a situated observer with distinctive priorities, purposes, and interpretive schemes).[82]

Social reality does not exist "out there," to be depicted either well or poorly by use of words, numbers, and code.[83] Rather, all these symbols have the capacity to make and remake the social world. Our linguistic capacity to describe and re-describe anew evinces a quintessentially human ambition to push the boundary of the "sayable," sometimes to come to consensus but also to sharpen differences. Charles Taylor speaks of "regestalt[ing] our experience" as particular forceful or telling phrasings, essays, poems, or novels resonate with us.[84] By seamlessly connecting science and art, description and self-expression, Taylor and Brown help us understand the role of literature and art in shaping a contemporary social theory fit to encompass and comment on our predicament. Rhetoric grounds our sense of social reality. It is a feature of social meanings crystallized into resonant language reinforcing sense with sound, inner feeling with an outward objective correlative.

Narratives about the future are drivers of the economy and society, and not simply post hoc efforts to understand what has already happened. This insight, going back to Keynes, has recently been recapitulated in Nobelist Robert Shiller's book *Narrative Economics*.[85] There is no physics of markets; economics is a human science, all the way down, and will be unpredictable as long as humans have free will. As Dennis Snower, president of the Kiel Institute for the World Economy, has observed, "The standard statistical analyses are no longer valid. They assume that we know the probabilities with which everything will occur. In reality, we are almost never in that position."[86] Narratives (and accessible versions of them) are critical to envisioning the future.[87] This is not necessarily a compromise of truth or validity. As law professor and social theorist Jack Balkin has argued, "Narrative memory structures are particularly useful for remembering what kinds of things are dangerous or advantageous, making complicated causal judgments about the future, determining what courses of action are helpful, recalling how to do things in a particular order, and learning and following social conventions."[88] Thinking deeply about how people make "complicated causal judgments" is a foundation of anticipatory social research, which intends not only to discern patterns from the past, but to shape the future.

As the trappings of objectivity fall away from standard, laissez-faire-promoting economics, there is more room for other ways of understanding commercial life to take root and thrive.[89] It is time to tell new stories about the nature and purpose of automation policy. Automated infrastructures should not exist merely to get us or our data from point A to point B at the lowest price; it should also be a source of fairly compensated, secure work. Our investments in robotics should not simply be a way to accumulate money for retirement; they must also be put to work to ensure there is a world we want to retire into. It is all too easy to waste resources in competition, so the third new law of robotics counsels in favor of scrutinizing many sectors for arms race dynamics.[90]

Advances in computation and machine learning have created sociotechnical systems that appear almost magical—or at least humanly inexplicable. Even the best engineers at Google or Facebook may not be able to reverse engineer how certain algorithms function in search results or news feeds. The acceleration of processing and inputs can make the digital world seem as unruly and uncontrollable as natural systems like weather. As the engineer and philosopher Mireille Hildebrandt has observed, we will increasingly find the digital environment as pervasively influential and consequential as the natural world.[91] The Technocene complements the Anthropocene, reminding us how consequential unintended consequences can become. Even if they never become formal legislation, the new laws of robotics can help us imagine futures where AI respects and helps preserve human dignity and identity.

AESTHETIC RESPONSE AND MORAL JUDGMENT

One of the most profound artistic exemplars of such an imagined future is the oeuvre of artist Ernesto Caivano. Elaborated in hundreds of drawings, paintings, and mixed media work, it is nothing less than an epic foretelling of a divergence of human and machine, which manages to maintain and respect the best qualities of each.

Though the digital has influenced the world of fine art and aesthetics in innumerable ways, few artists have responded with as ambitious and penetrating a body of work as Ernesto Caivano's. In an epic series of drawings entitled *After the Woods,* Caivano has told the story of lovers Versus and Polygon, whose courtship, separation, and eventual evolution allegorize fundamental tensions—and resonances—between nature and technology.[92] Joao Ribas, former curator of MIT's List Visual Art Center, has called the narrative an "amalgam of folklore, fairytale, and scientific speculation," which "serves as a search for meaning lost in our own abundance of information."[93]

The works alternate between the dramatic and the austerely formal—evoking layers of reality (from action to its data-driven representation) between nature, culture, and technology. They seamlessly move from a yearning romanticism, to an exultant exploration of alternate universes, to disciplined interrogations of what it means to use codes to represent perceptions. In a computationalist age—where leading thinkers present the universe as a computer, and powerful economists predict that the patterns of most persons' labor are readily translated via machine learning into software—Caivano patiently, wisely, builds a series of aesthetic responses to complement the moral judgments and political struggles that computation precipitates.

Caivano's work combines stylistic influences that are both contemporary and classic, from Albrecht Durer and the Flemish Renaissance, to modernist trends toward minimalism and abstraction in the work of Agnes Martin, to "art/science" hybrids (to use David Edwards's term for design thinking that blurs boundaries between realms of beauty and truth).[94] Caivano's *Echo Series* evokes the data exhaust we leave behind on the internet and that is increasingly sensed by robotics.[95] The data could be taken as a prelude to replacing us, but it is more justly apprehended as a mere reflection or trace of the embodied selves creating it. Similar themes animate Caivano's elegant *Philapores Navigating the Log and Code*. It portrays mythical birds flying through elements of a forest and its digital representation as a topology of geospatial data. The simultaneous depiction of map and thing mapped, reality and its image, is both jarring and inspiring. It is a symbolic evocation of so many lessons of responsible automation—for instance, of careful comparison between actual work and how it is coded. There is no simple didacticism here, but an imaginative freedom to rethink the relationship between the actual and its digital traces, disciplined by fidelity to natural forms. This work integrates aesthetics into science, and scientific concerns into art.

AUTOMATING THE AUTOMATORS

To the extent they reflect contemporary design thinking, Caivano's works fit comfortably with dominant digital aesthetics. They also convey a timelessness—of values that existed long before us and should endure long after—that is under pressure in an era of social acceleration.[96] The business maxim that work must be done ever faster is all too quickly instrumentalized as a rhetorical tool for suppressing workers' autonomy. American bosses, in their bid to demand more "flexibility" from a restive workforce, could once point to laborers abroad ready to take domestic workers' jobs; now, those bosses are prone to pointing to ever faster machines. Demanding more break time? A robot can work 24 / 7. Want higher wages? You only create incentives for the boss to replace you with software. Electricity and replacement parts are a lot cheaper than food and medicine.

Advances in software and robotics have been so rapid that they even raise questions about the replaceability of bosses themselves. Zappos has experimented with "holacracy," a flat-management style that leaves workers to self-organize their tasks.[97] The *Harvard Business Review* has repeatedly celebrated the "automation of management."[98] Apps like Uber's delegate management to a code layer, which connects riders with drivers. Bad drivers are no longer fired but "deactivated" by an algorithmic scoring tool.[99] Perhaps higher-level executives could find their own work ranked, rated, and ultimately completed by AI.

The idea of computerizing management may seem farcical, futuristic, or both. When a venture capital firm added an algorithm to its board of directors, asking it to vote up or down on companies based on a parsing of their prospectuses, the business press was skeptical.[100] The very conceivability of automated investment reveals something deeply wrong about our political economy. As time horizons shorten for chief executive officers, shareholders, and managers, their actions are becoming more stereotyped and algorithmic. Tactics such as tax inversions, share buybacks, offshoring, and substitution of machines for workers become a toolkit of tried-and-true heuristics. Managers can easily look back at how stock markets responded when other firms deployed similar tactics. So, too, is stacking the board that determines one's own compensation a tried, true, and repeated CEO tactic. A robot could do it.

Automating the automators—an idea often seized upon by left accelerationists—is less a revolutionary proposal to seize the means of production than it is an encapsulation of what is actually happening now in the world of finance and management. Robotization does not merely mean deploying a mechanized mannequin in place of a worker. It conveys standardization and repetition: one best way of doing a task is found and replicated. Managers may have many tools but habitually choose the same ends and the same means of achieving them. If mere routine or predictability renders their workers redundant, so too does it threaten the bosses.

Of course, in our daily lives, habit is a must. Routines dominate, because it's simply unnecessary to figure out—minutely—what's the best way to step out of bed, make a cup of coffee, or turn on the computer. But the human hankering for habit becomes pathological once writ large over space and time. This automatic way of thinking obscures better futures.

COCREATING THE FUTURE OF WORK

Today's cornucopianist automators are prone to view unions, regulators, cooperatives, and the professions as archaic speed bumps on a technocratic highway to abundance. In fact, each can democratize a future now largely dictated by technology and finance firms. We do well to democratize power beyond such firms, because, as roboticist Illah Reza Nourbakhsh has argued:

> Today most nonspecialists have little say in charting the role that robots will play in our lives. We are simply watching a new version of Star Wars scripted by research and business interests in real time, except that this script will become our actual world. . . . Familiar devices will become more aware, more interactive and more proactive; and entirely new robot creatures will share our spaces, public and private, physical and digital. . . . Eventually, we will need to read what they write, we will have to interact with them to conduct our business transactions, and we will often mediate our friendships through them.[101]

The terms of that cooperation may be monolithically neoliberal: to maximize investment returns for the wealthiest technologists and the most technologically advanced wealthy. Or they can be diverse, codeveloped with domain experts, and responsive to community values. Laissez-faire will reliably deliver us the former outcome; the four new laws of robotics promote the latter.

When the primary threat to legal order was the arbitrary decree of autocrats, judges and advocates resisted with the principle "a rule of law, not of men." As automation advances, we must now complement this adage with a commitment to "a rule of persons, not machines." Neither managers nor bureaucrats should be allowed to hide behind algorithmic modes of social ordering. Rather, personal responsibility for decisions is essential to maintaining legitimate governance, whether of state agencies or businesses.

We should also suspect technologists who assume that most jobs are predictable, capable of being learned by machines with access to enough data. This is a recipe for trapping society in the past—that is, in the "training set" from which machines will have learned. Automation enthusiasts may characterize the learning set as a mere starting point, enabling future autonomy from machines that learn from negative and positive stimuli in response to future actions. But the definition of "negative" and "positive" must itself be programmed in, as must opportunities to expand or contract their definitions. The organic development of moral responsibility (on the individual level) and self-governance (on a collective level) can only be attributed to AI itself via magical thinking familiar from fairy tales.

There is enormous power in contemporary computing. A rush to throw it at every major problem we have, untrammeled by laws, professions, or ethics, will be enormously tempting. To the extent there is a sociobiological human nature, its bent may well be toward arms races for relative position, accelerated by automation. As Simone Weil warned, "Not to exercise all the power at one's disposal is to endure the void. This is contrary to all the laws of nature. Grace alone can do it."[102]

The grace required will not be that of a cybernetic poet or a robotic angel of history.[103] There will be protracted political and legal disputes about the proper allocation of benefits and burdens. There will be numerous judgment calls about mundane topics—for example, when a physician can override a red alert on clinical-decision-support software, or a driver can switch off computer-assisted driving, or either gets to access the guts of computer systems now all too often opaque to them. Such disputes, far from being impediments to a better future, signal genuine and important conflicts over societal governance. We evade them at our peril.

Humane automation requires a wisdom of restraint. Widely read as parable of personal hubris, the tale of Icarus was also one of the first myths of technological overreach. In Ovid's telling, Daedalus and his son Icarus were stranded on an island. Daedalus constructed sets of wings for them from wax and feathers so they could fly to another shore. The daring plan worked, and father and son soared over the ocean. But Icarus ignored his father's warnings about flying too adventurously, too close to the sun. The wax melted, the wings disintegrated, and Icarus plunged to his death.

The myth of Icarus inspired other cautionary tales, such as Marlowe's *Faust*. The robotics field has its own version of the problem, known as the "uncanny valley"—the unease that a humanoid robot can cause when it soars beyond mere mechanism, to come very close to, but not quite recreate, human features, gestures, and ways of being in the world.[104] The term uncanny valley comes from a simple graphic, imagining a steady increase in robots' acceptability as they take on more functions and appearances of humans, until the machines become too humanoid—when their popularity suddenly crashes and is replaced by widespread aversion. However, the valley here is less cautionary tale than challenge. Ingenuity paired with gobs of data is supposed to catapult the clever robot designer out of the valley and toward machines as respected and welcomed as humans.

There is another approach, an alternate teleology. We can avoid the uncanny valley altogether by recognizing the limits of technology. In the haunting 1995 installation *Lovers,* the artist Teiji Furuhashi expressed this wisdom in a spare but immersive vision.[105] Projected onto four black scrims that surround the viewer in a darkened room, ghostly video of members of the Japanese performance collective Dumb Type shows them walking, standing, performing stylized movements of embrace or caution, running toward or away from one another, but never touching. Their images are often blurred by their movements, and there is no setting to distract from their bodies. Their motion may be triggered by the viewer's movements, but the connection is obscure. They emote energy and striving one moment, somber surrender the next. Created in the wake of the AIDS crisis that claimed Furuhashi's life, the work is at once a celebration of new technology's affordances, and a wise recognition of its limits. The dancer is not the dance.

Reflection on the culture inspired by automation is a step toward a common public language about the difficult political and personal choices AI will confront us with in coming decades. We can afford a future of robotics that is inclusive and democratic, reflecting the efforts and hopes of all persons as workers and citizens—not simply as consumers. We can embrace new forms of computation without abandoning quintessentially human roles to AI. Such a corrective course will consume much of the energy of current professions and will create new ones. It promises a positive and lasting freedom unattainable in conditions of technological drift.

NOTES

1. INTRODUCTION

1. Kevin Drum offers a useful overview of maximalist predictions, concluding that "no matter what job you name, robots will be able to do it." Kevin Drum, "You Will Lose Your Job to a Robot—and Sooner Than You Think," *Mother Jones,* November / December 2017, at https://www .motherjones.com/politics/2017/10/you-will-lose-your-job-to-a-robot-and -sooner-than-you-think/.

2. Several influential books have explored the economic implications of this narrative. Martin Ford, *Rise of the Robots: Technology and the Threat of a Jobless Future* (New York: Basic Books, 2015); Jerry Kaplan, *Humans Need Not Apply: A Guide to Wealth and Work in the Age of Artificial Intelligence* (New Haven: Yale University Press, 2015); Robin Hanson, *The Age of Em: Work, Love, and Life When Robots Rule the Earth* (Oxford: Oxford University Press, 2016).

3. Eric Brynjolffson and Andrew McAfee, *Race against the Machine: How the Digital Revolution Is Accelerating Innovation, Driving Productivity, and Irreversibly Transforming Employment and the Economy* (Lexington, MA: Digital Frontier Press, 2014).

4. Isaac Asimov, "Runaround," *Astounding Science Fiction,* March 1942. Asimov later introduced a meta-principle as his Zeroth Law of Robotics (namely, that robots must not harm humanity). Isaac Asimov, *Robots and Empire* (New York: Doubleday, 1985), 291.

5. Jack Balkin, "The Three Laws of Robotics in the Age of Big Data," *Ohio State Law Journal* 78 (2017): 1, 10. To his credit, Asimov was fully aware of these ambiguities, and he accentuated them to add drama to his stories.

6. These laws of robotics, like Asimov's, offer general guidance and are subject to exceptions and clarifications. As administrative lawyers are well aware, many statutes are vague and must be clarified (and reconciled with other, possibly conflicting laws) by agencies and courts.

7. With this focus, I follow in the footsteps of the second principle of robotics of the United Kingdom Engineering and Physical Sciences Research Council (UK EPSRC), which states that "humans, not robots, are responsible agents." Margaret Boden, Joanna Bryson, Darwin Caldwell, Kirsten Dautenhahn, Lillian Edwards, Sarah Kember, Paul Newman et al., Principles of Robotics (2010), https://epsrc.ukri.org /research/ourportfolio/themes/engineering/activities /principlesofrobotics/.

8. Government, civil society, business associations, and professional societies developed ethics guidelines for AI; for one meta-study of such guidelines, see Anna Jobin, Marcello Ienca, and Effy Vayena, "The Global Landscape of Ethics Guidelines," *Nature Machine Intelligence* 1, 389–399 (2019), https://doi.org/10.1038/s42256-019-0088-2. Regulatory proposals and guidances have also proliferated. According to the Fundamental Rights Agency of the European Union, there are over 250 initiatives of this nature within member states of the EU alone. Although the four laws of robotics proposed in this book cannot possibly do justice to these diverse documents, they do draw support from them, distilling key ideas in order to reflect widely shared values.

9. I focus on "robotics" and not "robots," to convey how embedded sensor / processor / actuator technology is in social systems.

10. Harold Wilensky, "The Professionalization of Everyone?," *American Journal of Sociology* 70, no. 2: 137–158 (describing the core of professional status as a "successful claim to exclusive technical competence and adherence to the service ideal.").

11. These legal protections are also reflected in a larger social commitment to complementary AI. The Federal Government of Germany, *Artificial Intelligence Strategy* 25 (2018), https://ec.europa.eu/knowledge4policy /publication/germany-artificial-intelligence-strategy_en ("The potential for AI to serve society as a whole lies in its promise of productivity gains going hand in hand with improvements for the workforce, delegating monotonous or dangerous tasks to machines so that human beings can focus on using their creativity to resolve problems.").

12. Lucas Mearian, "A. I. Guardian-Angel Vehicles Will Dominate Auto Industry, Says Toyota Exec," *Computerworld,* June 3, 2016.

13. Some forms of autopilot tend to de-skill pilots. Nicholas Carr, *The Glass Cage: How Our Computers Are Changing Us* (New York: Norton, 2015).

However, autopilots can also be designed to maintain or enhance the skills of pilots, preserving essential expertise. David Mindell, *Our Robots, Ourselves: Robotics and the Myths of Autonomy* (New York: Viking, 2015).

14. Thus, even firms at the forefront of the AI revolution, such as Microsoft, say that their goal "isn't to replace people with machines, but to supplement human capabilities with the unmatched ability of AI to analyze huge amounts of data and find patterns that would otherwise be impossible to detect." Microsoft, "The Future Computed: Artificial Intelligence and Its Role in Society" (2018), https://blogs.microsoft.com/wp-content/uploads/2018/02/The-Future-Computed_2.8.18.pdf.

15. Kate Crawford and Vladan Joler, *Anatomy of an AI System* (2018), https://anatomyof.ai/.

16. Information Technology Industry Council (ITI), AI Policy Principles Executive Summary 5 (2017), https://www.itic.org/public-policy/ITIAIPolicyPrinciplesFINAL.pdf.

17. The Agency for Digital Italy, *Artificial Intelligence at the Service of Citizens* 37, 54, 62 (2018), https://ia.italia.it/assets/whitepaper.pdf. See also Artificial Intelligence Industry Alliance (China), Joint Pledge on Artificial Intelligence Industry Self-Discipline (2019), https://www.newamerica.org/cybersecurity-initiative/digichina/blog/translation-chinese-ai-alliance-drafts-self-discipline-joint-pledge/), which holds that AI development should "prevent artificial intelligence from weakening and replacing humanity's position."

18. Frank Pasquale, "A Rule of Persons, Not Machines," *George Washington Law Review* 87, no. 1 (2019): 1–55.

19. Beijing Academy of AI, Beijing AI Principles (2019), https://www.baai.ac.cn/blog/beijing-ai-principles.

20. Daniella K. Citron and Robert Chesney, "Deep Fakes: A Looming Challenge for Privacy, Democracy, and National Security," *California Law Review* 107 (2019): 1753–1819.

21. European Group on Ethics in Science and New Technologies, European Commission, *Statement on Artificial Intelligence, Robotics and 'Autonomous' Systems* 16 (2018), https://publications.europa.eu/en/publication-detail/-/publication/dfebe62e-4ce9-11e8-be1d-01aa75ed71a1, which cites a fundamental principal of human dignity. See also IEEE, *Global Initiative on Ethics of Autonomous and Intelligent Systems, Ethically Aligned Design* 39 (1st. ed., 2019), https://standards.ieee.org/content/ieee

-standards/en/industry-connections/ec/autonomous-systems.html: "The duty to respect human dignity may require some limitations on the functions and capability of A / IS so that they do not completely replace humans, human functions, and / or 'human central thinking activities' such as judgment, discretion, and reasoning. . . . It may also involve preventing A / IS from deceiving or manipulating humans."

22. I borrow the term "mind children" from Moravec, even though my vision for the future of robotics is quite different. Hans Moravec, *Mind Children: The Future of Robot and Human Intelligence* (Cambridge, MA: Harvard University Press, 1990).

23. Even in such cases, a "burden of self-insurance" may be imposed. David C. Vladeck, "Machines without Principals: Liability Rules and Artificial Intelligence," *Washington Law Review* 89, no. 1 (2014): 117, 150.

24. The term "potentially responsible parties" is inspired by the Comprehensive Environmental Response, Compensation, and Liability Act, or CERCLA. 42 U.S.C. § 9607(a) (2012), which defines four categories of potentially liable parties.

25. Helena Horton, "Microsoft Deletes 'Teen Girl' AI after It Became a Hitler-Loving Sex Robot," *Telegraph,* March 24, 2017, https://www.telegraph.co.uk/technology/2016/03/24/microsofts-teen-girl-ai-turns-into-a-hitler-loving-sex-robot-wit/.

26. For the distinction between closed and open robotics, see M. Ryan Calo, "Open Robotics," *Maryland Law Review* 70 (2011): 571, 583–591; Diana Marina Cooper, "The Application of a 'Sufficiently and Selectively Open License' to Limit Liability and Ethical Concerns Associated with Open Robotics," in *Robot Law,* eds. Ryan Calo, A. Michael Froomkin, and Ian Kerr (Cheltenham, UK: Edward Elgar, 2016), 163, 164–165.

27. Staff of the US Securities and Exchange Commission and Staff of the US Commodity Futures Trading Commission, *Joint Study on the Feasibility of Mandating Algorithmic Descriptions for Derivatives* (April 2011), 16, 16n77, 24, https://www.sec.gov/news/studies/2011/719b-study.pdf.

28. John Markoff, "The Creature That Lives in Pittsburgh," *New York Times,* April 21, 1991, http://www.nytimes.com/1991/04/21/business/the-creature-that-lives-in-pittsburgh.html?pagewanted=all; Rodney Brooks, *Flesh and Machines: How Robots Will Change Us* (New York: Pantheon, 2002).

29. See John Markoff, *Machines of Loving Grace* (New York: HarperCollins, 2015), which describes pioneering IA work by Doug Engelbart and a long line of proteges. See also Doug Engelbart, *Augmenting Human Intellect: A Conceptual Framework* (Washington, DC: Air Force Office of Scientific Research, 1962).

30. Marc Andreessen, "Why Software Is Eating the World," *Wall Street Journal,* August 20, 2011, https://www.wsj.com/articles/SB10001424053111903480904576512250915629460.

31. Ryan Calo, "Robotics and the Lessons of Cyberlaw," *California Law Review* 103, no. 3 (2015): 513–563.

32. Ian Kerr, "Bots, Babes and the Californication of Commerce," *Ottawa Law and Technology Journal* 1 (2004): 285–325.

33. Rachel Lerman, "Be Wary of Robot Emotions; 'Simulated Love Is Never Love,'" *Phys.org,* April 26, 2019, https://phys.org/news/2019-04-wary-robot-emotions-simulated.html.

34. Natasha Dow Schüll, *Addiction by Design: Machine Gambling in Las Vegas* (Princeton: Princeton University Press, 2014); Ryan Calo, "Digital Market Manipulation," *George Washington Law Review* 82 (2014): 995; Neil Richards, "The Dangers of Surveillance," *Harvard Law Review* 126 (2019): 1934.

35. Mark Andrejevic, "Automating Surveillance," *Surveillance and Society* 17 (2019): 7.

36. Neil Selwyn, *Distrusting Educational Technology: Critical Questions for Changing Times* (New York: Routledge, 2014).

37. Laurence H. Tribe, *Channeling Technology through Law* (Chicago: Bracton, 1973).

38. Deborah G. Johnson, "The Role of Ethics in Science and Technology," *Cell* 29 (2010): 589–590; Deborah G. Johnson, "Software Agents, Anticipatory Ethics, and Accountability," in *The Growing Gap between Emerging Technologies and Legal-Ethical Oversight,* eds. Gary E. Marchant, Braden R. Allenby, and Joseph R. Herkert (New York: Spring, 2011), 61–76; Ari Ezra Waldman, *Privacy as Trust* (Oxford: Oxford University Press, 2018).

39. Mary Flanagan and Helen Nissenbaum, *Values at Play in Digital Games* (Cambridge, MA: MIT Press, 2014).

40. See, for example, Ann Cavoukian, "Privacy by Design: The 7 Foundational Principles," Office of the Information and Privacy Commissioner of Canada (2009).

41. For more on the privacy issues raised by robots and predictive analytics, see Drew Simshaw, Nicolas Terry, Kris Hauser, and M. L. Cummings, "Regulating Healthcare Robots: Maximizing Opportunities while Minimizing Risks," *Richmond Journal of Law and Technology* 27, no. 3 (2016): 1–38, 3. For an integration of law and design principles for effective notice, see Ari Ezra Waldman, "Privacy, Notice, and Design," *Stanford Technology Law Review* 21, no. 1 (2018): 129–184.

42. Hoofnagle calls this argumentative strategy the "denialists' deck of cards." Christopher Jay Hoofnagle, "The Denialists' Deck of Cards: An Illustrated Taxonomy of Rhetoric Used to Frustrate Consumer Protection Efforts," *ScienceBlogs*, February 9, 2007, https://scienceblogs.com/denialism/the-denialists-deck-of-cards.

43. Frank Pasquale, "Technology, Competition, and Values," *Minnesota Journal of Law, Science, and Technology* 8 (2007): 607–622; Peter Asaro, "*Jus Nascendi,* Robotic Weapons and the Martens Clause," in *Robot Law,* eds. Ryan Calo, A. Michael Froomkin, and Ian Kerr (Cheltenham, UK: Edward Elgar, 2016), 367–386.

44. Frank Pasquale, "Technology, Competition, and Values," *Minnesota Journal of Law, Science, and Technology* 8 (2007): 607–622

45. See, for example, Kenneth Anderson and Matthew C. Waxman, "Law and Ethics for Autonomous Weapon Systems: Why a Ban Won't Work and How the Laws of War Can," *Hoover Institution Stanford University Task Force on National Security and Law* (2013), responding to the Campaign to Stop Killer Robots.

46. P. W. Singer, *Wired for War: The Robotics Revolution and Conflict in the 21st Century* (New York: Penguin, 2009), 435.

47. Ray Kurzweil, *The Age of Spiritual Machines* (New York: Penguin, 1999).

48. Rebecca Crootof, "A Meaningful Floor for Meaningful Human Control," *Temple International and Comparative Law Journal* 30 (2016): 53–62; Paul Scharre, "Centaur Warfighting: The False Choice of Humans vs. Automation," *Temple International and Comparative Law Journal* 30 (2016): 151–166.

49.   Jeffrey L. Caton, *Autonomous Weapon Systems: A Brief Survey of Developmental, Operational, Legal, and Ethical Issues* (Carlisle, PA: US Army War College Press, 2015); Liang Qiao and Xiangsui Wang, *Unrestricted Warfare: China's Master Plan to Destroy America,* trans. Al Santoli (Panama City: Pan American Publishing, 2000).

50.   The projects of facial analysis mentioned in this paragraph are critiqued in Frank Pasquale, "When Machine Learning is Facially Invalid," *Communications of the ACM* 61, no. 9 (2018): 25–27.

51.   David Castlevecchi, "Can We Open the Black Box of AI?," *Scientific American,* October 5, 2016, https://www.scientificamerican.com/article /can-we-open-the-black-box-of-ai/.

52.   Jennifer Kavanagh and Michael D. Rich, *Truth Decay: An Initial Exploration of the Diminishing Role of Facts and Analysis in American Public Life,* Santa Monica, CA: RAND Corporation, 2018, https://www.rand.org /pubs/research_reports/RR2314.html; Alice Marwick and Rebecca Lewis, *Media Manipulation and Disinformation Online* (New York: Data and Society, 2017), https://datasociety.net/pubs/oh/DataAndSociety_MediaM anipulationAndDisinformationOnline.pdf.

53.   On the distinction between speech and conduct, see, for example, Claudia Haupt, "Professional Speech," *Yale Law Journal* 125, no. 5 (2016): 1238–1303.

54.   As Gil Eyal demonstrates, expertise may well exist "outside" the expert, not merely in sets of propositional knowledge, but also in institutions. Gil Eyal, *The Crisis of Expertise* (Medford, MA: Polity Press, 2019).

55.   Alex (Sandy) Pentland, *Honest Signals: How They Shape Our World* (Cambridge, MA: MIT Press, 2008), which discusses sociometric badges.

56.   See Andrew Abbott, *The System of Professions: An Essay on the Division of Expert Labor* (Chicago: University of Chicago Press, 2014); Eliot Freidson, *Professionalism, The Third Logic: On the Practice of Knowledge* (Chicago: University of Chicago Press, 2001).

57.   For a thoughtful perspective on the "automation of virtue," see Ian Kerr, "Digital Locks and the Automation of Virtue," in *From Radical Extremism to Balanced Copyright: Canadian Copyright and the Digital Agenda,* ed. Michael Geist (Toronto: Irwin Law, 2010), 247–303.

58.   Hope Reese, "Updated: Autonomous Driving Levels 0 to 5: Understanding the Differences," *TechRepublic,* January 20, 2016, http://www

.techrepublic.com/article/autonomous-driving-levels-0-to-5-understanding-the-differences/.

59.   Cathy O'Neil, *Weapons of Math Destruction: How Big Data Increases Inequality and Threatens Democracy* (New York: Crown, 2016); Frank Pasquale, *The Black Box Society: The Secret Algorithms That Control Money and Information* (Cambridge, MA: Harvard University Press, 2015); danah boyd and Kate Crawford, "Critical Questions for Big Data: Provocations for a Cultural, Technological, and Scholarly Phenomenon." *Information, Communication, and Society* 15, no. 5 (2012): 662–679.

60.   Bryant Walker Smith, "How Governments Can Promote Automated Driving," *New Mexico Law Review* 47 (2017): 99–138.

61.   Ibid., 114.

62.   I give only the obvious examples here. For a more imaginative survey of future work, largely focused on adaptations to new technology, see Thomas Frey, "55 Jobs of the Future," *Futurist Speaker,* November 11, 2011, http://www.futuristspeaker.com/business-trends/55-jobs-of-the-future/.

63.   US Department of Transportation, Federal Automated Vehicles Policy: Accelerating the Next Revolution in Roadway Safety (September 2016), 9, https://www.transportation.gov/AV/federal-automated-vehicles-policy-september-2016.

64.   Rules governing liability here will be critical. Marc Canellas and Rachel Haga, "Unsafe at Any Level: The U.S. NHTSA's Levels of Automation Are a Liability for Automated Vehicles," *Communications of the ACM* 63, no. 3 (2020): 31–34.

65.   Nor are the strictly technological challenges easy to address. Roberto Baldwin, "Self-Driving Cars are Taking Longer to Build Than Everyone Thought," *Car and Driver,* May 10, 2020, https://www.caranddriver.com/features/a32266303/self-driving-cars-are-taking-longer-to-build-than-everyone-thought/.

66.   AI Now Institute, *AI Now 2019 Report,* December 2019, 8, 45–47, https://ainowinstitute.org/AI_Now_2019_Report.pdf.

67.   Henry Mance, "Britain Has Had Enough of Experts, says Gove," *Financial Times,* June 3, 2016, https://www.ft.com/content/3be49734-29cb-11e6-83e4-abc22d5d108c.

68. Gil Eyal, *The Crisis of Expertise* (Cambridge, UK: Polity Press, 2019), 20.

69. General Raymond Thomas, head of the US Special Operations Command, recalled the chairman of Google, Eric Schmidt, saying this to him in July 2016. Kate Conger and Cade Metz, "'I Could Solve Most of Your Problems': Eric Schmidt's Pentagon Offensive," *New York Times*, May 2, 2020, https://www.nytimes.com/2020/05/02/technology/eric-schmidt -pentagon-google.html.

70. Will Davies, "Elite Power under Advanced Neoliberalism," *Theory, Culture and Society* 34, nos. 5–6 (2017): 233 (italics in original).

71. I believe this integration of democratic and expertocratic values may be the key to resolving the divide between "diggers" and "levelers" in Science and Technology Studies, trenchantly developed by Philip Mirowski in a recent draft. Philip Mirowski, "Democracy, Expertise and the Post-truth Era: An Inquiry into the Contemporary Politics of STS," (working paper, version 1.1, April 2020), https://www.academia.edu/42682483 /Democracy_Expertise_and_the_Post-Truth_Era_An_Inquiry_into_the _Contemporary_Politics_of_STS.

72. Hubert L. Dreyfus, *What Computers Still Can't Do* (Cambridge, MA: MIT Press, 1992).

73. Gert P. Westert, Stef Groenewoud, John E. Wennberg, Catherine Gerard, Phil DaSilva, Femke Atsma, and David C. Goodman, "Medical Practice Variation: Public Reporting a First Necessary Step to Spark Change," *International Journal for Quality in Health Care* 30, no. 9 (2018): 731—735, https://doi.org/10.1093/intqhc/mzy092.

74. Anna B. Laakmann, "When Should Physicians Be Liable for Innovation?," *Cardozo Law Review* 36 (2016): 913–968.

75. Andy Kiersz, "These Are the Industries Most Likely to Be Taken Over by Robots," *World Economic Forum,* April 25, 2019, https://www.weforum .org/agenda/2019/04/these-are-the-industries-most-likely-to-be-taken -over-by-robots; Andrew Berg, Edward F. Buffie, and Luis-Felipe Zanna, "Should We Fear the Robot Revolution? (The Correct Answer Is Yes)," IMF Working Paper WP/18/116, May 21, 2018, https://www.imf.org/en /Publications/WP/Issues/2018/05/21/Should-We-Fear-the-Robot -Revolution-The-Correct-Answer-is-Yes-44923.

76. Pedro Domingos, *The Master Algorithm: How the Quest for the Ultimate Learning Machine Will Remake Our World* (New York: Basic Books, 2015).

77. Hugo Duncan, "Robots to Steal 15 Million of Your Jobs, Says Bank Chief: Doom-Laden Carney Warns Middle Classes Will Be 'Hollowed Out' by New Technology," *Daily Mail*, December 5, 2008, http://www.dailymail .co.uk/news/article-4003756/Robots-steal-15m-jobs-says-bank-chief -Doom-laden-Carney-warns-middle-classes-hollowed-new-technology .html#ixzz4SDCt2Pql.

78. See, for example, Clayton M. Christensen, Clayton Christensen, Curtis W. Johnson, and Michael B. Horn, *Disrupting Class* (New York: McGraw-Hill, 2008) and Clayton M. Christensen, Jerome Grossman, and Jason Hwang, *The Innovator's Prescription: A Disruptive Solution for Health Care* (New York: McGraw-Hill, 2009), which examines disruptive innovations in health care.

79. Kenneth Scheve and David Stasavage, *Taxing the Rich: A History of Fiscal Fairness in the United States and Europe* (Princeton: Princeton University Press, 2016).

80. Alondra Nelson, "Society after Pandemic," *Items: Insights from the Social Sciences,* April 23, 2020, at https://items.ssrc.org/covid-19-and-the-social -sciences/society-after-pandemic/. John Urry has also argued that "the terrain of future studies should be reclaimed for social science and, in a way, for people in their day to day lives." Urry, *What Is the Future?* (Malden: Polity, 2016).

81. Joseph Weizenbaum, *Computer Power and Human Reason: From Judgment to Calculation* (New York: Freeman, 1976).

82. Aaron Smith and Monica Anderson, *Automation in Everyday Life* (Washington, DC: Pew Research Center, October 4, 2017), https://www .pewinternet.org/2017/10/04/automation-in-everyday-life/.

2. HEALING HUMANS

1. Credit to Richard K. Morgan, *Altered Carbon* (New York: Ballantine, 2003) for the vision of bodies as "sleeves" for the mind. The general theme of rapid medical advance ending or long-delaying aging is common in medical futurism. Aubrey De Grey, *Ending Aging: The Rejuvenation Breakthroughs That Could Reverse Human Aging in Our Lifetime* (New York: St. Martin's, 2008).

2. For an excellent overview of the law and policy issues raised by even this more realistic vision, see Ian Kerr and Jason Millar, "Robots and Artificial

Intelligence in Healthcare" in *Canadian Health Law & Policy,* eds. Joanna Erdman, Vanessa Gruben and Erin Nelson (Ottawa: Lexis Nexis: 2017), 257–280.

3. The data deficiencies of many health care systems were dramatically demonstrated during the COVID pandemic of 2020, when even many wealthy nations lacked the testing infrastructure to understand the scope and severity of the problem.

4. Sarah L. Cartwright and Mark P. Knudson, "Evaluation of Acute Abdominal Pain in Adults," *American Family Physician* 77 (2008): 971–978.

5. Sharifa Ezat Wan Puteh and Yasmin Almualm, "Catastrophic Health Expenditure among Developing Countries," *Health Systems Policy and Research* 4, no. 1 (2017), doi:10.21767/2254-9137.100069; Daniel Callahan and Angela A. Wasunna, "The Market in Developing Countries: An Ongoing Experiment," in *Medicine and the Market: Equity v. Choice* (Baltimore: The Johns Hopkins University Press, 2006), 117.

6. Veronica Pinchin, "I'm Feeling Yucky: Searching for Symptoms on Google," *Keyword,* June 20, 2016, https://googleblog.blogspot.com/2016 /06/im-feeling-yucky-searching-for-symptoms.html.

7. Kelly Reller, "Mayo Assists Google in Providing Quick, Accurate Symptom and Condition Information," *Mayo Clinic,* June 21, 2016, http://newsnetwork.mayoclinic.org/discussion/mayo-clinic-assists -google-in-providing-quick-accurate-symptom-and-related-condition -information/.

8. Ian Steadman, "IBM's Watson Is Better at Diagnosing Cancer than Human Doctors," *Wired,* February 11, 2013, http://www.wired.co.uk /article/ibm-watson-medical-doctor.

9. Ajay Agrawal, Joshua Gans, and Avi Goldfarb, *Prediction Machines: The Simple Economics of Artificial Intelligence* (Cambridge, MA: Harvard Business Review Press, 2018).

10. For example, a self-driving car's "vision" system may interpret a stop sign as a "45 miles per hour" sign if some pieces of tape are placed on the sign. Kevin Eykholt, Ivan Evtimov, Earlence Fernandes, Bo Li, Amir Rahmati, Chaowei Xiao, Atul Prakash, et al., "Robust Physical-World Attacks on Deep Learning Visual Classification," arXiv:1707.08945v5 [cs.CR] (2018).

11. Eric Topol, *Deep Medicine: How Artificial Intelligence Can Make Healthcare Human Again* (New York: Basic Books, 2019).

12. Kim Saverno, "Ability of Pharmacy Clinical Decision-Support Software to Alert Users about Clinically Important Drug-Drug Interactions," *Journal of American Medical Informatics Association* 18, no. 1 (2011): 32–37.

13. Lorenzo Moja, "Effectiveness of Computerized Decision Support Systems Linked to Electronic Health Records: A Systematic Review and Meta-Analysis," *American Journal of Public Health* 104 (2014): e12–e22; Mariusz Tybinski, Pavlo Lyovkin, Veronika Sniegirova, and Daniel Kopec, "Medical Errors and Their Prevention," *Health* 4 (2012): 165–172.

14. Committee on Patient Safety and Health Information Technology Board on Health Care Services, *Health IT and Patient Safety: Building Safer Systems for Better Care* (Washington, DC: The National Academies Press, 2012), 39.

15. Lorenzo Moja, Koren Hyogene Kwag, Theodore Lytras, Lorenzo Bertizzolo, Linn Brandt, Valentina Pecoraro et al., "Effectiveness of Computerized Decision Support Systems Linked to Electronic Health Records: A Systematic Review and Meta-Analysis," *American Journal of Public Health* 104 (2014): e12–e22. See also Elizabeth Murphy, "Clinical Decision Support: Effectiveness in Improving Quality Processes and Clinical Outcomes and Factors that May Influence Success," *Yale Journal of Biology and Medicine* 87 (2014): 187–197, which showed that there was reduced risk in certain situations like deep vein thrombosis and pulmonary embolisms by 41 percent.

16. For example, with respect to subsidies, in the United States, certain subsidized providers must use a CDSS that checks for drug-drug or drug-allergy interactions. See, for example, Health Information Technology for Clinical and Economic Health Act, 42 U.S.C. § 300jj(13) (2009), which stipulates that "qualified electronic health record[s]" must be able to "provide clinical decision support"; see also 45 C.F.R. § 170.314 (2015); Medicare and Medicaid Programs; Electronic Health Record Incentive Program—Stage 3 and Modifications to Meaningful Use in 2015 through 2017, 80 Fed. Reg. 62761, 62838 (October 16, 2015).

17. M. Susan Ridgely and Michael D. Greenberg, "Too Many Alerts, Too Much Liability: Sorting through the Malpractice Implications of Drug-Drug Interaction Clinical Decision Support," *St. Louis University Journal of Health Law & Policy* 5, no. 2 (2012): 257, 274. Regulators may eventually want to examine the cases of "physicians with inappropriately

high override rates" to identify whether there may be a problem with the CDSS or with the physician's judgment. Karen C. Nanji, Sarah P. Slight, Diane L. Seger, Insook Cho, Julie M. Fiskio, Lisa M. Redden, Lynn A. Volk, and David W. Bates, "Overrides Of Medication-Related Clinical Decision Support Alerts in Outpatients," *Journal of American Medical Informatics Association* 21, no. 3 (2014): 487–491.

18. A. Michael Froomkin, Ian Kerr, and Joelle Pineau, "When AIs Outperform Doctors: Confronting the Challenges of a Tort-Induced Over-Reliance on Machine Learning," *Arizona Law Review* 61 (2019): 33–100.

19. In the case of facial recognition, there is a well-documented failure of AI systems to recognize the faces of persons of color. Joy Buolamwini, "Actionable Auditing: Coordinated Bias Disclosure Study," MIT Civic Media Project, https://www.media.mit.edu/projects/actionable-auditing-coordinated-bias-disclosure-study/overview/. These failures parallel racial disparities in US health care. Dayna Bowen Matthew, *Just Medicine: A Cure for Racial Inequality in American Health Care* (New York: New York University Press, 2015).

20. Ziad Obermeyer, Brian Powers, Christine Vogeli, and Sendhil Mullainathan, "Dissecting Racial Bias in an Algorithm Used to Manage the Health of Populations," *Science* 366 no. 6464 (2019): 447–453; Ruha Benjamin, "Assessing Risk, Automating Racism," *Science* 366, no. 6464 (2019): 421–422. There are many ways to reduce such disparities. Frank Pasquale and Danielle Keats Citron, "Promoting Innovation while Preventing Discrimination: Policy Goals for the Scored Society," *Washington Law Review* 89 (2014): 1413–1424.

21. Adewole S. Adamson and Avery Smith, "Machine Learning and Health Care Disparities in Dermatology," *JAMA Dermatology* 154, no. 11 (2018): 1247. This lack of diversity also afflicts genomics research. Alice B. Popejoy, Deborah I. Ritter, Kristy Crooks, Erin Currey, Stephanie M. Fullerton, Lucia A. Hindorff, Barbara Koenig, et al., "The Clinical Imperative for Inclusivity: Race, Ethnicity, and Ancestry (REA) in Genomics," *Human Mutation* 39, no. 11 (2018): 1713–1720.

22. Tom Baker, *The Medical Malpractice Myth* (Chicago: University of Chicago Press, 2005); Alex Stein, "Toward a Theory of Medical Malpractice," *Iowa Law Review* 97 (2012): 1201–1258.

23. Meredith Broussard, *Artificial Unintelligence: How Computers Misunderstand the World* (Cambridge, MA: MIT Press, 2018), 32.

24. LeighAnne Olsen, J. Michael McGuinnis, and Dara Alsner, eds., *Learning Healthcare System: Workshop Summary* (Washington, DC: Institute of Medicine, 2007).

25. See, for example, Eric Topol, *Deep Medicine: How Artificial Intelligence Can Make Healthcare Human Again* (New York: Basic Books, 2019), which cites concerns about "cherry-picking results or lack of reproducibility"; Matthew Zook, Solon Barocas, danah boyd, Kate Crawford, Emily Keller, Seeta Peña Gangadharan, Alyssa Goodman et al., "10 Simple Rules for Responsible Big Data Research," *PLoS Computational Biology* 13, no. 3 (2017): e1005399, identifying similar limits; danah boyd and Kate Crawford, "Critical Questions for Big Data," *Journal of Information, Communication and Society* 15 (2012): 662–679.

26. Carolyn Criado Perez, *Invisible Women: Data Bias in a World Designed for Men* (New York: Abrams, 2019).

27. Ibid., 199–200.

28. Jack Balkin, "The Three Laws of Robotics in the Age of Big Data," *Ohio State Law Journal* 78 (2017): 1, 40.

29. Nathan Cortez, "The Mobile Health Revolution?," *University of California at Davis Law Review* 47 (2013): 1173–1230.

30. Wendy Wagner, "When All Else Fails: Regulating Risky Products through Tort Regulation," *Georgetown Law Journal* 95 (2007): 693, 694.

31. Wagner, "When All Else Fails."

32. John Villasenor, *Products Liability and Driverless Cars: Issues and Guiding Principles for Legislation* (Brookings Institution, 2014), https://www.brookings.edu/wp-content/uploads/2016/06/Products _Liability_and_Driverless_Cars.pdf.

33. Martha T. McCluskey, "Defining the Economic Pie, Not Dividing or Maximizing It," *Critical Analysis of Law* 5, no. 1 (2018): 77–98; Mariana Mazzucato, *The Value of Everything: Making and Taking in the Global Economy* (New York: Hachette, 2018).

34. See, for example, *Tunkl v. Regents of the University of California,* 383 P.2d 441 (Cal. 1963) and its progeny.

35. Margaret Jane Radin, *Boilerplate: The Fine Print, Vanishing Rights, and the Rule of Law* (Princeton: Princeton University Press, 2012).

36. Passing costs to insurers has been raised by several commentators. See, for example, Jacob Turner, *Robot Rules: Regulating Artificial Intelligence* (Cham, Switzerland: Palgrave MacMillan, 2019), 115.

37. National Academy of Sciences, *Best Care at Lower Cost: The Path to Continuously Learning Health Care in America* (Washington, DC: National Academies Press, 2013).

38. Frank Pasquale, "Health Information Law," in *The Oxford Handbook of U.S. Health Law,* eds. I. Glenn Cohen, Allison Hoffman, and William M. Sage (New York: Oxford University Press, 2016), 193–212; Sharona Hoffman, *Electronic Health Records and Medical Big Data: Law and Policy* (New York: Cambridge University Press, 2016).

39. Steven E. Dilsizian and Eliot L. Siegel, "Artificial Intelligence in Medicine and Cardiac Imaging: Harnessing Big Data and Advanced Computing to Provide Personalized Medical Diagnosis and Treatment," *Current Cardiology Reports* 16 (2014): 441, 445.

40. Melissa F. Hale, Mark E Mcalindon, and Reena Sidhu, "Capsule Endoscopy: Current Practice and Future Directions," *World Journal of Gastroenterology* 20 (2014): 7752.

41. James H. Thrall, "Data Mining, Knowledge Creation, and Work Process Enhancement in the Second Generation of Radiology's Digital Age," *Journal of the American College of Radiology* 10 (2013): 161–162.

42. Simon Head, *Mindless: Why Smarter Machines Are Making Dumber Humans* (New York: Basic Books, 2014).

43. Nina Bernstein, "Job Prospects Are Dimming for Radiology Trainees," *New York Times,* March 27, 2013, http://www.nytimes.com/2013/03/28 /health/trainees-in-radiology-and-other-specialties-see-dream-jobs -disappearing.html?pagewanted=all&_r=1&; European Society of Radiology, "The Consequences of the Economic Crisis in Radiology," *Insights into Imaging* 6, no. 6 (2015): 573–577.

44. Giles W. L. Boland, "Teleradiology Coming of Age: Winners and Losers," *American Journal of Roentgenology* 190 (2008): 1161.

45. The relevant policymakers have focused on cutting costs, rather than investing in advances. For an insightful look at the changing economics of radiology, see Frank Levy and Max P. Rosen, "How Radiologists Are Paid: An Economic History, Part IV: End of the Bubble," *Journal of the American College of Radiology* (2020), https://doi.org/10.1016/j.jacr.2020.02.016.

46. Sharona Hoffman, "The Drugs Stop Here," *Food and Drug Law Journal* 67, no. 1 (2012): 1.

47. Gerd Gigerenzer, Wolfgang Gaissmaier, Elke Kurz-Milcke, Lisa M. Schwartz, and Steven Woloshin, "Helping Doctors and Patients Make

Sense of Health Statistics," *Psychological Science in the Public Interest* 8 (2008): 53. This paper criticizes physicians' lack of statistical literacy, particularly relating to the "number needed to treat" (NNT) figure related to medicine.

48. Geoff White, "Child Advice Chatbots Fail to Spot Sexual Abuse," *BBC News,* December 11, 2018, https://www.bbc.com/news/technology -46507900.

49. Dinesh Bhugra, Allan Tasman, Sounitra Pathare, Stefan Prebe, Shubalade Smith, John Torous, Melissa Arbuckle et al., "The WPA Lancet Psychiatry Commission on the future of Psychiatry," *The Lancet Psychiatry* 4, no. 10 (2017):775–818.

50. For an excellent overview of such privacy concerns, see Piers Gooding, "Mapping the Rise of Digital Mental Health Technologies: Emerging Issues for Law and Society," *International Journal of Law and Psychiatry* 67 (2019): 1–11. A key problem in the US is that "unlike licensed professionals who provide traditional mental health services, chatbots are not subject to confidentiality obligations." Scott Stiefel, "The Chatbot Will See You Now: Protecting Mental Health Confidentiality in Software Applications," *Columbia Science and Technology Law Review* 20 (2019): 333–387.

51. For example, Australian researchers concluded that "commercial actors, including powerful app distributors and commercial third parties were rarely subjects of policy initiatives, despite having considerable power to affect app user outcomes." Lisa Parker, Lisa Bero, Donna Gillies, Melissa Raven, and Quinn Grundy, "The 'Hot Potato' of Mental Health App Regulation: A Critical Case Study of the Australian Policy Arena," *International Journal of Health Policy and Management* 8 (2019): 168–176.

52. Simon Leigh and Steve Flatt, "App-Based Psychological Interventions: Friend or Foe?" *Evidence Based Mental Health* 18 (2015): 97–99.

53. Jack M. Balkin, "Information Fiduciaries and the First Amendment," *UC Davis Law Review* 49 (2016): 1183–1234.

54. Maurice E. Stucke and Ariel Ezrachi, "Alexa et al., What Are You Doing with My Data?" *Critical Analysis of Law* 5 (2018): 148–169; Theodore Lee, "Recommendations for Regulating Software-Based Medical Treatments: Learning from Therapies for Psychiatric Conditions," *Food and Drug Law Journal* 73 (2018): 66–102. For a good general treatment of trust and privacy, see Ari Ezra Waldman, *Privacy as Trust* (Cambridge, UK: Cambridge University Press, 2018).

55. Daniel Carpenter, *Reputation and Power: Organizational Image and Pharmaceutical Regulation at the FDA* (Princeton: Princeton University Press, 2010).

56. Jennifer Akre Hill, "Creating Balance: Problems with DSHEA and Suggestions for Reform," *Journal of Food Law & Policy* 2 (2006): 361–396; Peter J. Cohen, "Science, Politics, and the Regulation of Dietary Supplements: It's Time to Repeal DSHEA," *American Journal of Law & Medicine* 31 (2005): 175–214.

57. Joseph Weizenbaum, *Computer Power and Human Reason: From Judgment to Calculation* (San Francisco: W. H. Freeman, 1976).

58. Enrico Gnaulati, *Saving Talk Therapy: How Health Insurers, Big Pharma, and Slanted Science Are Ruining Good Mental Health Care* (Boston: Beacon Press, 2018).

59. Elizabeth Cotton, "Working in the Therapy Factory," *Healthcare: Counselling and Psychotherapy Journal* 20, no. 1 (2020): 16–18; Catherine Jackson and Rosemary Rizq, eds., The Industrialisation of Care: Counselling, Psychotherapy and the Impact of IAPT (Monmouth, UK: PCCS Books, 2019).

60. Peter Kramer, *Against Depression* (New York: Viking, 2005).

61. See, for example, Johann Hari, *Lost Connections: Uncovering the Real Causes of Depression* (New York: Bloomsbury, 2018).

62. Frank A. Pasquale, "Privacy, Autonomy, and Internet Platforms," in *Privacy in the Modern Age: The Search for Solutions,* ed. Marc Rotenberg, Julia Horwitz, and Jeramie Scott (New York: New Press, 2015), 165–173.

63. Employment law regimes matter, too. French patterns of protecting worker interests are a better foundation for health interventions in the workplace than comparable US law. Julie C. Suk, "Preventive Health at Work: A Comparative Approach," *American Journal of Comparative Law* 59, no. 4 (2011): 1089-1134.

64. Natasha Dow Schüll, *Addiction by Design: Machine Gambling in Law Vegas* (Princeton: Princeton University Press, 2014).

65. Gordon Hull and Frank Pasquale, "Toward a Critical Theory of Corporate Wellness," *Biosocieties* 13 (2017): 190–212; William Davies, *The Happiness Industry: How the Government and Big Business Sold Us Well-Being* (London: Verso Books, 2016).

66. Martha Nussbaum, *Upheavals of Thought: The Intelligence of Emotions* (New York: Cambridge University Press, 2001), 22.

67. Davies, *The Happiness Industry.*

68. Dayna Matthew, *Just Medicine: A Cure for Racial Inequality in American Health Care* (New York: New York University Press, 2015).

69. Frank Pasquale, "Professional Judgment in an Era of Artificial Intelligence," *boundary2* 46 (2019): 73–101.

70. Brett Frischmann and Evan Selinger, *Re-engineering Humanity* (New York: Cambridge University Press, 2018).

71. Larissa MacFarquhar, "A Tender Hand in the Presence of Death," *New Yorker,* July 11, 2016, http://www.newyorker.com/magazine/2016/07/11/the-work-of-a-hospice-nurse.

72. Nellie Bowles, "Human Contact Is Now a Luxury Good," *New York Times,* March 23, 2019, https://www.nytimes.com/2019/03/23/sunday-review/human-contact-luxury-screens.html.

73. GeriJoy, "Care.Coach Intro Video," *YouTube,* February 2, 2019, https://www.youtube.com/watch?v=GSTHIG4vx_0, at 0:50.

74. Kathy Donaghie, "My Robot Companion Has Changed My Life," *Independent IE,* April 15, 2019, https://www.independent.ie/life/health-wellbeing/health-features/my-robot-companion-has-changed-my-life-38009121.html.

75. Byron Reeves and Clifford Nass, *The Media Equation: How People Treat Computers, Television, and New Media Like Real People and Places* (New York: Cambridge University Press, 1996), 252.

76. Kate Darling, "'Who's Johnny?' Anthropomorphic Framing in Human-Robot Interaction, Integration, and Policy," in *Robot Ethics 2.0: From Autonomous Cars to Artificial Intelligence,* eds. Patrick Lin, Ryan Jenkins, and Keith Abney (New York: Oxford University Press, 2017), 173–192.

77. This is particularly true when it comes to potential recording, and reporting, of the elder's behavior. Ari Ezra Waldman, *Privacy as Trust* (New York: Cambridge University Press, 2018).

78. "Green Nursing Homes," PBS, *Religion and Ethics Newsweekly,* July 20, 1997, http://www.pbs.org/wnet/religionandethics/2007/07/20/july-20-2007-green-house-nursing-homes/3124/.

79. Sherry Turkle, "A Nascent Robotics Culture: New Complicities for Companionship," *AAAI Technical Report Series,* July 2006, 1, http://web.mit.edu/~sturkle/www/nascentroboticsculture.pdf.

80. Aviva Rutkin, "Why Granny's Only Robot Will Be a Sex Robot," *New Scientist,* July 8, 2016, https://www.newscientist.com/article/2096530-why-grannys-only-robot-will-be-a-sex-robot/.

81. Ai-Jen Poo, *The Age of Dignity: Preparing for the Elder Boom in a Changing America* (New York: The New Press, 2015), 92–99.

82. See, for example, Berry Ritholtz, "Having Trouble Hiring? Try Paying More," *Bloomberg Opinion,* September 8, 2016, https://www.bloomberg.com/view/articles/2016-09-08/having-trouble-hiring-try-paying-more; Thijs van Rens, "Paying Skilled Workers More Would Create More Skilled Workers," *Harvard Business Review,* May 19, 2016, https://hbr.org/2016/05/paying-skilled-workers-more-would-create-more-skilled-workers.

83. See, for example, Swiss Business Hub Japan, *Healthcare Tech in Japan: A Booming Market,* 2018, https://swissbiz.jp/wp-content/uploads/2018/01/sge_healthcaretech_japan_infographic.pdf.

84. Joi Ito, "Why Westerners Fear Robots and the Japanese Do Not," *Wired,* July 30, 2018, https://www.wired.com/story/ideas-joi-ito-robot-overlords.

85. "Japan Is Both Obsessed With and Resistant to Robots," *Economist,* November, 2018, https://www.economist.com/asia/2018/11/08/japan-is-both-obsessed-with-and-resistant-to-robots.

86. Selina Cheng, "'An Insult to Life Itself': Hayao Miyazaki Critiques an Animation Made by Artificial Intelligence," *Quartz,* December 10, 2016, https://qz.com/859454/the-director-of-spirited-away-says-animation-made-by-artificial-intelligence-is-an-insult-to-life-itself/.

87. Madhavi Sunder, "Cultural Dissent," *Stanford Law Review* 54 (2001): 495–568.

88. Amartya Sen, "Human Rights and Asian Values," Sixteenth Annual Morgenthau Memorial Lecture on Ethics and Foreign Policy, May 25, 1997, Carnegie Council for Ethics in International Rights, https://www.carnegiecouncil.org/publications/archive/morgenthau/254. https://papers.ssrn.com/sol3/papers.cfm?abstract_id=304619/.

89. Danit Gal, "Perspectives and Approaches in AI Ethics: East Asia," in *The Oxford Handbook of Ethics of Artificial Intelligence*, eds. Markus Dubber, Frank Pasquale, and Sunit Das (Oxford: Oxford University Press, 2020), chapter 32.

90. Martha Fineman, "The Vulnerable Subject and the Responsive State," *Emory Law Journal* 60 (2010): 251–276.

91.   Huei-Chuan Sung, Shu-Min Chang, Mau-Yu Chin, and Wen-Li Lee, "Robot-Assisted Therapy for Improving Social Interactions and Activity Participation among Institutionalized Older Adults: A Pilot Study," *Asia Pacific Psychiatry* 7 (2015): 1–6, https://onlinelibrary.wiley.com/doi/epdf/10.1111/appy.12131.

92.   Nina Jøranson, Ingeborg Pedersen, Anne Marie Rokstad, Geir Aamodt, Christine Olsen, and Camilla Ihlebæk, "Group Activity with Paro in Nursing Homes: Systematic Investigation of Behaviors in Participants," *International Psychogeriatrics* 28 (2016): 1345–1354, https://doi.org/10.1017/S1041610216000120.

93.   Aimee van Wynsberghe, "Designing Robots for Care: Care Centered Value-Sensitive Design," *Science and Engineering Ethics* 19, (2013): 407–433.

94.   *Ik ben Alice* [Alice Cares], directed and written by Sander Burger (Amsterdam: KeyDocs, 2015), DCP, 80 min.

95.   European Commission, "Special Eurobarometer 382: Public Attitudes Towards Robots," September 2012, http://ec.europa.eu/public_opinion/archives/ebs/ebs_382_en.pdf.

96.   Eliot L. Freidson, *Professionalism: The Third Logic* (Chicago: University of Chicago Press, 2001).

97.   The "general wellness" category is defined and explored in U.S. Food and Drug Administration, General Wellness: Policy for Low Risk Devices (2019), at https://www.fda.gov/media/90652/download.

98.   Jennifer Nicholas et al., "Mobile Apps for Bipolar Disorder: A Systematic Review of Features and Content Quality," *Journal of Medical Internet Research* 17, no. 8 (2015): e198.

99.   Vinayak K. Prasad and Adam S. Cifu, *Ending Medical Reversal: Improving Outcomes, Saving Lives* (Baltimore: Johns Hopkins University Press, 2015).

3. BEYOND MACHINE LEARNERS

1.   Xue Yujie, "Camera above the Classroom: Chinese Schools Are Using Facial Recognition on Students. But Should They?" *Sixth Tone,* March 26, 2019, http://www.sixthtone.com/news/1003759/camera-above-the-classroom.

2.   Ibid. See also Louise Moon, "Pay Attention at the Back: Chinese School Installs Facial Recognition Cameras to Keep An Eye on Pupils," *South China Morning Post,* May 16, 2018, https://www.scmp.com/news/china /society/article/2146387/pay-attention-back-chinese-school-installs-facial -recognition.

3.   "Notice of the State Council Issuing the New Generation of Artificial Intelligence Development Plan," State Council Document [2017] No. 35, trans. Flora Sapio, Weiming Chen, and Adrian Lo, Foundation for Law and International Affairs, https://flia.org/wp-content/uploads/2017/07/A -New-Generation-of-Artificial-Intelligence-Development-Plan-1.pdf.

4.   Melissa Korn, "Imagine Discovering That Your Teaching Assistant Really Is a Robot," *Wall Street Journal,* May 6, 2016, http://www.wsj.com/articles /if-your-teacher-sounds-like-a-robot-you-might-be-on-to-something -1462546621.

5.   IBM has made similar points about Watson's role relative to lawyers and doctors, primarily as a tool of the existing professionals rather than as machine to replace them.

6.   Tamara Lewin, "After Setbacks, Online Courses Are Rethought," *New York Times,* December 10, 2016, https://www.nytimes.com/2013/12/11/us /after-setbacks-online-courses-are-rethought.html.

7.   Neil Selwyn, *Distrusting Education Technology: Critical Questions for Changing Times* (New York: Routledge 2014), 125.

8.   David F. Labaree, "Public Goods, Private Goods: The American Struggle over Educational Goals," *American Educational Research Journal* 34, no. 1 (Spring 1997): 39, 46; Danielle Allen, "What is Education For?," *Boston Review,* May 9, 2016, http://bostonreview.net/forum/danielle-allen -what-education.

9.   US Department of Education, "Education Department Releases College Scorecard to Help Students Choose Best College for Them," press release, February 13, 2013, https://www.ed.gov/news/press-releases/education -department-releases-college-scorecard-help-students-choose-best-college -them.

10. Claudia D. Goldin and Lawrence F. Katz, *The Race Between Education and Technology* (Cambridge, MA: Harvard University Press, 2009).

11. Charles Taylor, *Human Agency and Language: Philosophical Papers*, vol. 1 (Cambridge, UK: Cambridge University Press, 1985).

12. Jane Mansbridge, "Everyday Talk in the Deliberative System," in *Deliberative Politics: Essays on Democracy and Disagreement,* ed. by Stephen Macedo (New York: Oxford University Press, 1999), 1–211.

13. Matthew Arnold, *Culture and Anarchy* (New York: Oxford University Press 2006).

14. Nikhil Goyal, *Schools on Trial: How Freedom and Creativity Can Fix Our Educational Malpractice* (New York: Doubleday, 2016).

15. Audrey Watters, "The Monsters of Education Technology," *Hack Education* (blog), December 1, 2014, http://hackeducation.com/2014/12/01/the-monsters-of-education-technology.

16. Stephen Petrina, "Sidney Pressey and the Automation of Education 1924–1934," *Technology and Culture* 45 (2004): 305–330.

17. Sidney Pressey, "A Simple Apparatus Which Gives Tests and Scores—And Teaches," *School and Society* 23 (1926): 373–376.

18. B.F. Skinner, *The Technology of Teaching,* e-book ed. (1968; repr., Cambridge, MA: B.F. Skinner Foundation, 2003), PDF.

19. Audrey Watters, "Teaching Machines," *Hack Education* (blog), April 26, 2018, http://hackeducation.com/2018/04/26/cuny-gc.

20. Natasha Dow Schüll, *Addiction by Design* (Princeton: Princeton University Press, 2012).

21. Rowan Tulloch and Holly Eva Katherine Randell-Moon, "The Politics of Gamification: Education, Neoliberalism and the Knowledge Economy," *Review of Education, Pedagogy, and Cultural Studies* 40, no. 3 (2018): 204–226.

22. Tristan Harris, "How Technology Is Hijacking Your Mind—From a Magician and Google Design Ethicist," *Medium,* May 18, 2016, https://medium.com/swlh/how-technology-hijacks-peoples-minds-from-a-magician-and-google-s-design-ethicist-56d62ef5edf3#.ryse2c3rl.

23. Kate Darling, "Extending Legal Protection to Social Robots: The Effects of Anthropomorphism, Empathy and Violent Behavior towards Robotic Objects," in *Robot Law,* ed. Ryan Calo, A. Michael Froomkin, and Ian Kerr (Northampton, MA: Edward Elgar, 2016), 213–233.

24. Hashimoto Takuya, Naoki Kato, and Hiroshi Kobayashi, "Development of Educational System with the Android Robot SAYA and Evaluation," *International Journal of Advanced Robotic Systems* 8, no. 3 (2011): 51–61.

25. Ibid., 52.

26. Ibid., 60.

27. Hashimoto Takuya, Hiroshi Kobayashi, Alex Polishuk, and Igor Verner, "Elementary Science Lesson Delivered by Robot," *Proceedings of the 8th ACM / IEEE International Conference on Human-Robot Interaction* (March 2013): 133–134; Patricia Alves-Oliveira, Tiago Ribeiro, Sofia Petisca, Eugenio di Tullio, Francisco S. Melo, and Ana Paiva, "An Empathic Robotic Tutor for School Classrooms: Considering Expectation and Satisfaction of Children as End-Users," in *Social Robotics: International Conference on Social Robotics,* eds. Adriana Tapus, Elisabeth André, Jean-Claude Martin, François Ferland, and Mehdi Ammi (Cham, Switzerland: Springer, 2015), 21–30.

28. Neil Selwyn, *Should Robots Replace Teachers?* (Medford, MA: Polity Press, 2019).

29. Bureau of Labor Statistics, "Employment Characteristics of Families Summary," news release, April 18, 2019, https://www.bls.gov/news.release /famee.nr0.htm. Sixty-three percent of families with children under eighteen had both parents employed in 2018.

30. Noel Sharkey and Amanda Sharkey, "The Crying Shame of Robot Nannies: An Ethical Appraisal," *Interaction Studies* 11 (2010): 161–163.

31. Amanda J. C. Sharkey, "Should We Welcome Robot Teachers?" *Ethics and Information Technology* 18 (2016): 283–297. For thoughtful reflection on the potential confusion of those interacting with bots, see Ian Kerr, "Bots, Babes and the Californication of Commerce" *Ottawa Law and Technology Journal* 1 (2004): 285–325, discussing an avatar called ElleGirlBuddy.

32. This education of the sentiments toward accepting AI personhood is, in my view, the key to understanding Turing's classic "test": not as a philosophical standard for assessing the success of AI, but as a rhetorical device to acclimate readers to assume that that the status of personhood must be granted upon the performance of certain task(s).

33. This is important, because, as lawyer and philosopher Mireille Hilde-brandt has argued, the "data-driven agency" common in algorithmic systems builds on "information and behaviour, not meaning and action." Mireille Hildebrandt, "Law as Information in the Era of Data-Driven Agency," *Modern Law Review* 79 (2016): 1, 2.

34.   Chapter 8 offers an extended discussion of this distinction.

35.   Kate Darling, "Extending Legal Protection to Social Robots: The Effects of Anthropomorphism, Empathy, and Violent Behavior towards Robotic Objects," in *Robot Law,* eds. Ryan Calo, A. Michael Froomkin, and Ian Kerr (Cheltenham, UK: Edward Elgar, 2016),

36.   Benjamin Herold, "Pearson Tested 'Social-Psychological' Messages in Learning Software, with Mixed Results," *Education Week: Digital Education* (blog), April 17, 2018, http://blogs.edweek.org/edweek /DigitalEducation/2018/04/pearson_growth_mindset_software.html.

37.   See, for example, James Grimmelmann, "The Law and Ethics of Experiments on Social Media Users," *Colorado Technology Law Journal* 13 (2015): 219–227, which discusses letters written by Grimmelmann and Leslie Meltzer Henry on Facebook emotional experimentation; Chris Gilliard, "How Ed Tech Is Exploiting Students," *Chronicle of Higher Education,* April 8, 2018, https://www.chronicle.com/article/How-Ed -Tech-Is-Exploiting/243020.

38.   Sarah Schwarz, "YouTube Accused of Targeting Children with Ads, Violating Federal Privacy Law," *Education Week: Digital Education* (blog), April 13, 2018, http://blogs.edweek.org/edweek/DigitalEducation /2018/04/youtube_targeted_ads_coppa_complaint.html; John Montgallo, "Android App Tracking Improperly Follows Children, Study," *QR Code Press,* April 18, 2018, http://www.qrcodepress.com/android-app-tracking -improperly-follows-children-study/8534453/.

39.   James Bridle, "Something Is Wrong on the Internet," *Medium,* November 6, 2017, https://medium.com/@jamesbridle/something-is-wrong -on-the-internet-c39c471271d2.

40.   Nick Statt, "YouTube Will Reportedly Release a Kids' App Curated by Humans," *Verge,* April 6, 2018, https://www.theverge.com/2018/4/6/17208532 /youtube-kids-non-algorithmic-version-whitelisted-conspiracy-theories.

41.   Natasha Singer, "How Companies Scour Our Digital Lives for Clues to Our Health," *New York Times,* February 25, 2018, https://www.nytimes .com/2018/02/25/technology/smartphones-mental-health.html.

42.   Sam Levin, "Facebook Told Advertisers It Can Identify Teens Feeling 'Insecure' and 'Worthless,'" *Guardian,* May 1, 2017, https://www .theguardian.com/technology/2017/may/01/facebook-advertising-data -insecure-teens.

43. Julie E. Cohen, "The Regulatory State in the Information Age," *Theoretical Inquiries in Law* 17, no. 2 (2016): 369–414.

44. Erving Goffman, *The Presentation of Self in Everyday Life* (New York: Anchor, 1959).

45. For an example of an Australian family court case in which the court refused to accept evidence derived from a recording device in a child's toy, see *Gorman & Huffman* [2015] FamCAFC 127.

46. Complaint and Request for Investigation, Injunction, and Other Relief at 2, In re: Genesis Toys and Nuance Communications, submitted by the Electronic Privacy Information Center, the Campaign for a Commercial Free Childhood, the Center for Digital Democracy, and Consumers Union (December 6, 2016), https://epic.org/privacy/kids/EPIC-IPR-FTC-Genesis-Complaint.pdf. Some European authorities intervened creatively and forcefully. The Norwegian Consumer Council dramatized surveillance dolls' powers in a public service video. The German Federal Network Agency advised parents to destroy them, since they violated laws against concealed spying devices.

47. Ibid., 10.

48. "Millions of Voiceprints Quietly Being Harvested as Latest Identification Tool," *Guardian,* October 13, 2014, https://www.theguardian.com/technology/2014/oct/13/millions-of-voiceprints-quietly-being-harvested-as-latest-identification-tool.

49. Nicola Davis, "'High Social Cost' Adults Can Be Predicted from as Young as Three, Says Study," *Guardian,* December 12, 2016, https://www.theguardian.com/science/2016/dec/12/high-social-cost-adults-can-be-identified-from-as-young-as-three-says-study; Avshalom Caspi, Renate M. Houts, Daniel W. Belsky, Honalee Harrington, Sean Hogan, Sandhya Ramrakha, Richie Poulton, and Terrie E. Moffitt, "Childhood Forecasting of a Small Segment of the Population with Large Economic Burden," *Nature Human Behaviour* 1, no. 1 (2017): article no. UNSP 0005.

50. Mary Shacklett, "How Artificial Intelligence Is Taking Call Centers to the Next Level," *Tech Pro Research,* June 12, 2017, http://www.techproresearch.com/article/how-artificial-intelligence-is-taking-call-centers-to-the-next-level/.

51. For an extraordinary chronicling of the problems generated by edtech, see Audrey Watters, "The 100 Worst Ed-Tech Debacles of the Decade,"

Hacked Education (blog), http://hackeducation.com/2019/12/31/what-a
-shitshow.

52. Scott R. Peppet, "Unraveling Privacy: The Personal Prospectus and the
Threat of a Full-Disclosure Future," *Northwestern University Law Review*
105 (2011): 1153, 1201.

53. Lisa Feldman Barrett, Ralph Adolphs, Stacy Marsella, Aleix M. Martinez,
and Seth D. Pollak, "Emotional Expressions Reconsidered: Challenges to
Inferring Emotion from Human Facial Movements," *Psychological
Science in the Public Interest* 20, no. 1 (2019), https://journals.sagepub
.com/stoken/default+domain/10.1177%2F1529100619832930-FREE/pdf.

54. Kate Crawford, Roel Dobbe, Theodora Dryer, Genevieve Fried, Ben
Green, Elizabeth Kaziunas, Amba Kak, et al., *AI Now 2019 Report* (New
York: AI Now Institute, 2019), https://ainowinstitute.org/AI_Now_2019
_Report.html.

55. Ben Williamson, Sian Bayne, and Suellen Shayet, "The datafication of
teaching in Higher Education: critical issues and perspectives," *Teaching
in Higher Education* 25, no. 4 (2020): 351–365.

56. Hannah Arendt, *Between Past and Future* (New York: Penguin, 1954):
"The responsibility for the development of the child turns in a certain sense
against the world: the child requires special protection and care so that
nothing destructive may happen to him from the world. But the world,
too, needs protection to keep it from being overrun and destroyed by the
onslaught of the new that bursts upon it with each new generation" (182).

57. Seymour Papert, *Mindstorms: Children, Computers and Powerful Ideas*,
2nd ed. (New York: Basic Books, 1993), 5.

58. Cathy O'Neil, *Weapons of Math Destruction: How Big Data Increases
Inequality and Threatens Democracy* (New York: Crown, 2016), 8.

59. Jacqueline M. Kory, Sooyeon Jeong, and Cynthia Breazeal, "Robotic
Learning Companions for Early Language Development," *Proceedings of
the 15th ACM on International Conference on Multimodal Interaction*
(2013), 71–72, https://dam-prod.media.mit.edu/x/files/wp-content
/uploads/sites/14/2015/01/KoryJeongBreazeal-ICMI-13.pdf.

60. Jacqueline M. Kory and Cynthia Breazeal, "Storytelling with Robots:
Learning Companion for Preschool Children's Language Development,"
*Robot and Human Interactive Communication RO-MAN, The 23rd IEEE
International Symposium on IEEE* (2014), http://www.jakory.com/static
/papers/kory-storybot-roman-v1-revisionA.pdf.

61. Deanna Hood, Severin Lemaignan, and Pierre Dillenbourg, "When Children Teach a Robot to Write: An Autonomous Teachable Humanoid Which Uses Simulated Handwriting," *Proceedings of the Tenth Annual ACM / IEEE International Conference on Human-Robot Interaction ACM* (2015), 83–90, https://infoscience.epfl.ch/record/204890/files/hood2015 when.pdf.

62. Fisher Price, "Think & Learn Code-a-pillar," https://www.fisher-price .com/en-us/product/think-learn-code-a-pillar-twist-gfp25; KinderLab Robotics, "Kibo," http://kinderlabrobotics.com/kibo/; Nathan Olivares-Giles, "Toys That Teach the Basics of Coding," *Wall Street Journal,* August 20, 2015, https://www.wsj.com/articles/toys-that-teach-the -basics-of-coding-1440093255.

63. Darling, "Extending Legal Protection to Social Robots."

64. A. Michael Froomkin and P. Zak Colangelo, "Self-Defense against Robots and Drones," *Connecticut Law Review* 48 (2015): 1–70.

65. Evgeny Morozov, *To Save Everything, Click Here: The Folly of Techno-logical Solutionism* (New York: Public Affairs, 2013).

66. Sherry Turkle, "A Nascent Robotics Culture: New Complicities for Companionship," in *Annual Editions: Computers in Society 10 / 11,* ed. Paul De Palma, 16th ed. (New York: McGraw-Hill, 2010), chapter 37.

67. Margot Kaminski, "Robots in the Home: What Will We Have Agreed to?," *Idaho Law Review* 51 (2015): 661–678; Woodrow Hartzog, "Unfair and Deceptive Robots," *Maryland Law Review* 74 (2015): 785–832; Joanna J. Bryson, "The Meaning of the EPSRC Principles of Robotics," *Connection Science* 29, no. 2 (2017): 130–136.

68. Neda Atanasoski and Kalindi Vora, *Surrogate Humanity: Race, Robots, and the Politics of Technological Futures* (Chapel Hill: Duke University Press, 2019).

69. Kentaro Toyama, *Geek Heresy: Rescuing Social Change from the Cult of Technology* (New York: Public Affairs, 2015).

70. Paul Prinsloo offers a nuanced and insightful analysis of global power differentials' effects on the datafication of education. Paul Prinsloo, "Data Frontiers and Frontiers of Power in (Higher) Education: A View of/from the Global South," *Teaching in Higher Education: Critical Perspectives* 25, no. 4 (2019): 366–383.

71. Andrew Brooks, "The Hidden Trade in Our Second-Hand Clothes Given to Charity," *Guardian,* February 13, 2015, https://www.theguardian.com

/sustainable-business/sustainable-fashion-blog/2015/feb/13/second-hand
-clothes-charity-donations-africa.

72. Shoshana Zuboff, "The Secrets of Surveillance Capitalism," *Frankfurter Allegemeine Zeitung,* March 5, 2016, http://www.faz.net/aktuell/feuilleton /debatten/the-digital-debate/shoshana-zuboff-secrets-of-surveillance -capitalism-14103616-p2.html.

73. Steven Rosenfeld, "Online Public Schools Are a Disaster, Admits Billionaire, Charter School-Promoter Walton Family Foundation," *AlterNet,* February 6, 2016, http://www.alternet.org/education/online-public-schools -are-disaster-admits-billionaire-charter-school-promoter-walton; Credo Center for Research on Education Outcomes, *Online Charter School Study 2015* (Stanford, CA: Center for Research on Education Outcomes, 2015), https://credo.stanford.edu/pdfs/OnlineCharterStudyFinal2015.pdf.

74. Audrey Watters, "Top Ed-Tech Trends of 2016: The Business of Education Technology," *Hack Education* (blog), December 5, 2016, http://2016trends .hackeducation.com/business.html; "Eschools Say They Will Appeal Audits Determining Inflated Attendance," *Columbus Dispatch,* October 4, 2016, http://www.dispatch.com/news/20161003/eschools-say-they-will-appeal -audits-determining-inflated-attendance/1; Benjamin Herold, "Problems with For-Profit Management of Pa. Cybers," *Education Week,* November 3, 2016, http://www.edweek.org/ew/articles/2016/11/03/problems-with-for -profit-management-of-pa-cybers.html; Benjamin Herold, "A Virtual Mess: Inside Colorado's Largest Online Charter School," *Education Week,* November 3, 2016, http://www.edweek.org/ew/articles/2016/11/03/a-virtual -mess-colorados-largest-cyber-charter.html; Erin McIntyre, "Dismal Performance by Idaho Virtual Charters Result in 20% Grad Rate," *Education Dive,* January 29, 2016, http://www.educationdive.com/news/dismal -performance-by-idaho-virtual-charters-result-in-20-grad-rate/412945/.

75. Watters, "Top Ed-Tech Trends of 2016." For an update of ever more bad ideas, see Audrey Watters, "The 100 Worst Ed-Tech Debacles of the Decade," Hack Education, December 31, 2019, http://hackeducation.com /2019/12/31/what-a-shitshow.

76. Watters, "The Business of Education Technology."

77. Donald T. Campbell, "Assessing the Impact of Planned Social Change," in *Social Research and Public Policies,* ed. Gene M. Lyons (Hanover, NH: University Press of New England, 1975), 35; Brian Kernighan, "We're Number One!" *Daily Princetonian*, October 25, 2010.

78. Watters, "The Business of Education Technology."

79. Lilly Irani, "Justice for 'Data Janitors,'" Public Books, January 15, 2015, http://www.publicbooks.org/nonfiction/justice-for-data-janitors; Trebor Scholz, *Introduction to Digital Labor: The Internet as Playground and Factory* (New York: Routledge, 2013), 1.

80. Phil McCausland, "A Rural School Turns to Digital Education. Is It a Savior or Devil's Bargain?" *NBC News*, May 28, 2018, https://www.nbcnews.com/news/us-news/rural-school-turns-digital-education-it-savior-or-devil-s-n877806.

81. Douglas Rushkoff, *Program or be Programmed: Ten Commands for a Digital Age* (New York: OR Books, 2010), 7–8.

82. Ibid., 9.

83. Ofcom, "Children and Parents: Media Uses and Attitudes Report: 2018," January 29, 2019, 11, https://www.ofcom.org.uk/__data/assets/pdf_file/0024/134907/children-and-parents-media-use-and-attitudes-2018.pdf; Gwenn Schurgin O'Keeffe and Kathleen Clarke-Pearson, "The Impact of Social Media on Children," *Pediatrics* 127, no. 4 (2011): 800, 802.

84. Senator Bernie Sanders has proposed a plan that would aid colleges in such a digital transition by directing funds away from sports and toward "activities that improve instructional quality and academic outcomes." College for All Act of 2019, S.1947, 116th Cong. (2019).

85. Todd E. Vachon and Josef (Kuo-Hsun) Ma, "Bargaining for Success: Examining the Relationship between Teacher Unions and Student Achievement," *Sociological Forum* 30 (2015): 391, 397–399. Michael Godsey argues that "there is a profound difference between a local expert teacher using the Internet and all its resources to supplement and improve his or her lessons, and a teacher facilitating the educational plans of massive organizations. Why isn't this line being publicly and sharply delineated, or even generally discussed?" Godsey, "The Deconstruction of the K–12 Teacher: When Kids Can Get Their Lessons from the Internet, What's Left for Classroom Instructors to Do?," *Atlantic,* March 25, 2015, https://www.theatlantic.com/education/archive/2015/03/the-deconstruction-of-the-k-12-teacher/388631/.

86. For further descriptions of special commissions proposed to address new technology, see Oren Bracha and Frank Pasquale, "Federal Search Commission? Access, Fairness, and Accountability in the Law of Search," *Cornell Law Review* 93 (2008): 1149; Ryan Calo, "The Case for a Federal

Robotics Commission," *Center for Technology Innovation at Brookings,* September 15, 2014, https://www.brookings.edu/wp-content/uploads /2014/09/RoboticsCommissionR2_Calo.pdf. One recent inspiration for such work is Elizabeth Warren, "Unsafe at Any Rate," *Democracy Journal,* Summer 2007, no. 5, https://democracyjournal.org/magazine/5/unsafe-at -any-rate/.

4. THE ALIEN INTELLIGENCE OF AUTOMATED MEDIA

1. Hartmut Rosa, *Social Acceleration: A New Theory of Modernity* (New York: Columbia University Press, 2015).

2. For a chart showing US newspaper advertising revenue from 1950 to 2015, see Mark J. Perry, "US Newspaper Advertising Revenue: Adjusted for Inflation, 1950 to 2015," *American Enterprise Institute: Carpe Diem,* June 16, 2016, http://www.aei.org/publication/thursday-night-links-10/. For a study showing how much of online ad revenue has been diverted away from media and toward digital intermediaries, see Pricewater- houseCoopers, *ISBA Programmatic Supply Chain Transparency Study* (2020).

3. Anthony Nadler, Matthew Crain, and Joan Donovan, *Weaponizing the Digital Influence Machine: The Political Perils of Online Ad Tech* (New York: Data & Society, 2018), https://datasociety.net/output/weaponizing -the-digital-influence-machine/; Whitney Phillips, *The Oxygen of Amplifi- cation: Better Practices for Reporting on Extremists, Antagonists, and Manipulators* (New York: Data & Society, 2018), https://datasociety.net/wp -content/uploads/2018/05/FULLREPORT_Oxygen_of_Amplification_DS .pdf; Martin Moore, *Democracy Hacked: Political Turmoil and Information Warfare in the Digital Age* (London: Oneworld, 2018).

4. Philip N. Howard, *Lie Machines: How to Save Democracy from Troll Armies, Deceitful Robots, Junk News Operations, and Political Operatives* (Oxford: Oxford University Press, 2020); Siva Vaidhyanathan, *Anti-Social Media: How Facebook Disconnects Us and Undermines Democracy* (Oxford: Oxford University Press, 2018).

5. Bence Kollanyi, Philip N. Howard, and Samuel C. Woolley, "Bots and Automation over Twitter During the Second U.S. Presidential Debate," *COMPROP Data Memo,* October 19, 2016, http://politicalbots.org /wp-content/uploads/2016/10/Data-Memo-Second-Presidential -Debate.pdf.

6.   Mark Andrejevic, *Automated Media* (New York: Routledge, 2019), 62.

7.   The current relationship between journalists and AI is expertly analyzed in Nicholas Diakopolous, *Automating the News: How Algorithms Are Rewriting the Media* (Cambridge, MA: Harvard University Press, 2019).

8.   The reforms I am proposing will also help most other content creators seeking online audiences, including entertainers. This chapter focuses on media automation.

9.   Amanda Taub and Max Fisher, "Where Countries Are Tinderboxes and Facebook Is a Match," *New York Times,* April 21, 2018, https://www .nytimes.com/2018/04/21/world/asia/facebook-sri-lanka-riots.html; Damian Tambini, "Fake News: Public Policy Responses," London School of Economics Media Policy Brief No. 20, March 2017, http://eprints.lse.ac .uk/73015/1/LSE%20MPP%20Policy%20Brief%2020%20-%20Fake%20 news_final.pdf.

10.   Stephan Russ-Mohl, "Bots, Lies and Propaganda: The New Misinforma-tion Economy," *European Journalism Observatory,* October 20, 2016, https://en.ejo.ch/latest-stories/bots-lies-and-propaganda-the-new -misinformation-economy; Carole Cadwalladr, "Facebook's Role in Brexit—and the Threat to Democracy," TED talk, filmed April 2019 at TED2019 Conference, 15:16, https://www.ted.com/talks/carole _cadwalladr_facebook_s_role_in_brexit_and_the_threat_to _democracy/transcript?language=en.

11.   Alex Shepard, "Facebook Has a Genocide Problem," *New Republic,* March 15, 2018, https://newrepublic.com/article/147486/facebook -genocide-problem; Euan McKirdy, "When Facebook becomes 'The Beast': Myanmar Activists Say Social Media Aids Genocide," *CNN,* April 6, 2018, https://www.cnn.com/2018/04/06/asia/myanmar-facebook -social-media-genocide-intl/index.html. For a searing critique of internet giants' complicity in sowing hatred, see Mary Ann Franks, *The Cult of the Constitution* (Palo Alto, CA: Stanford University Press, 2019).

12.   Michael H. Keller, "The Flourishing Business of Fake YouTube Views," *New York Times,* August 11, 2018, at https://www.nytimes.com /interactive/2018/08/11/technology/youtube-fake-view-sellers.html.

13.   Zeynep Tufekci, "Facebook's Ad Scandal Isn't a 'Fail,' It's a Feature," *New York Times,* September 23, 2017, https://www.nytimes.com/2017/09/23 /opinion/sunday/facebook-ad-scandal.html; Zeynep Tufekci, "YouTube, the Great Radicalizer," *New York Times,* March 10, 2018, https://www

.nytimes.com/2018/03/10/opinion/sunday/youtube-politics-radical.html; Siva Vaidhyanathan, *The Googlization of Everything (and Why We Should Worry)* (Berkeley: University of California Press, 2011); Siva Vaidhyanathan, *Antisocial Media: How Facebook Disconnects Us and Undermines Democracy* (New York: Oxford University Press, 2018).

14. Evelyn Douek, "Facebook's 'Oversight Board': Move Fast with Stable Infrastructure and Humility," *North Carolina Journal of Law & Technology* 21 (2019): 1–78.

15. Frank Pasquale, "Platform Neutrality: Enhancing Freedom of Expression in Spheres of Private Power," *Theoretical Inquiries in Law* 17 (2017): 480, 487. For a critique of this situation, see Olivier Sylvain, "Intermediary Design Duties," *Connecticut Law Review* 50 (2018): 203–278; Ari Ezra Waldman, "Durkheim's Internet: Social and Political Theory in Online Society," *New York University Journal of Law and Liberty* 7 (2013): 345, 373–379.

16. See Carole Cadwalladr, "How to Bump Holocaust Deniers off Google's Top Spot?" *Guardian,* December 17, 2016, https://www.theguardian.com /technology/2016/dec/17/holocaust-deniers-google-search-top-spot.

17. Sydney Schaedel, "Did the Pope Endorse Trump?" *FactCheck.org,* October 24, 2016, http://www.factcheck.org/2016/10/did-the-pope -endorse-trump/; Dan Evon, "Spirit Cooking," *Snopes,* November 5, 2016, http://www.snopes.com/john-podesta-spirit-cooking/.

18. Ian Bogost and Alexis C. Madrigal, "How Facebook Works for Trump," *Atlantic,* April 18, 2020, at https://www.theatlantic.com/technology /archive/2020/04/how-facebooks-ad-technology-helps-trump-win /606403/. For YouTube analysis by engineer Guillaume Chaslot, see Paul Lewis, "'Fiction is Outperforming Reality': How YouTube's Algorithm Distorts Truth," *Guardian*, February 2, 2018, https://www.theguardian .com/technology/2018/feb/02/how-youtubes-algorithm-distorts-truth.

19. Kyle Chayka, "Facebook and Google Make Lies As Pretty As Truth," *Verge,* December 6, 2016, http://www.theverge.com/2016/12/6/13850230 /fake-news-sites-google-search-facebook-instant-articles; Janet Guyon, "In Sri Lanka, Facebook Is Like the Ministry of Truth," *Quartz*, April 22, 2018, https://qz.com/1259010/how-facebook-rumors-led-to-real-life -violence-in-sri-lanka/.

20. Eric Lubbers, "There Is No Such Thing as the Denver Guardian, Despite that Facebook Post You Just Saw," *Denver Post,* November 5, 2016, http://www.denverpost.com/2016/11/05/there-is-no-such-thing-as-the -denver-guardian/.

21. Brett Molina, "Report: Fake Election News Performed Better Than Real News on Facebook," *USA Today,* November 17, 2016, http://www.usatoday.com/story/tech/news/2016/11/17/report-fake-election-news-performed-better-than-real-news-facebook/94028370/.

22. Timothy B. Lee, "The Top 20 Fake News Stories Outperformed Real News at the End of the 2016 Campaign," *Vox,* November 16, 2016, https://www.vox.com/new-money/2016/11/16/13659840/facebook-fake-news-chart.

23. Joel Winston, "How the Trump Campaign Built an Identity Database and Used Facebook Ads to Win the Election," *Medium*, November 18, 2016, https://medium.com/startup-grind/how-the-trump-campaign-built-an-identity-database-and-used-facebook-ads-to-win-the-election-4ff7d24269ac#.4oaz94q5a.

24. Paul Lewis, "'Utterly Horrifying': Ex-Facebook Insider Says Covert Data Harvesting Was Routine," *Guardian,* March 20, 2018, https://www.theguardian.com/news/2018/mar/20/facebook-data-cambridge-analytica-sandy-parakilas; Nicol Perlroth and Sheera Frenkel, "The End for Facebook's Security Evangelist," *New York Times,* March 20, 2018, https://www.nytimes.com/2018/03/20/technology/alex-stamos-facebook-security.html.

25. Zeynep Tufekci, "Mark Zuckerberg Is in Denial," *New York Times,* November 15, 2016, http://www.nytimes.com/2016/11/15/opinion/mark-zuckerberg-is-in-denial.html?_r=2.

26. Olivier Sylvain, "Intermediary Design Duties," *Connecticut Law Review* 50, no. 1 (2018): 203; Danielle Keats Citron and Mary Anne Franks, "The Internet as a Speech Machine and Other Myths Confounding Section 230 Reform" (working paper, Public Law Research Paper No. 20-8, Boston University School of Law, Massachusetts, 2020); Carrie Goldberg, *Nobody's Victim: Fighting Psychos, Stalkers, Pervs, and Trolls* (New York: Plume, 2019), 38.

27. Craig Silverman, "Facebook Is Turning to Fact-Checkers to Fight Fake News," *BuzzFeed News,* December 15, 2016, https://www.buzzfeed.com/craigsilverman/facebook-and-fact-checkers-fight-fake-news?utm_term=.phQ1y0OexV#.sn4XJ8ZoLA.

28. Timothy B. Lee, "Facebook Should Crush Fake News the Way Google Crushed Spammy Content Farms," *Vox,* December 8, 2016, http://www.vox.com/new-money/2016/12/8/13875960/facebook-fake-news-google.

29. Safiya Umoja Noble, *Algorithms of Oppression: How Search Engines Reinforce Racism* (New York: New York University Press, 2018); Carole Cadwalladr, "Google, Democracy and the Truth about Internet Search,"

Guardian, December 4, 2016, https://www.theguardian.com/technology/2016/dec/04/google-democracy-truth-internet-search-facebook.

30. Manoel Horta Ribeiro, Raphael Ottoni, Robert West, Virgílio A. F. Almeida, and Wagner Meira Jr., "Auditing Radicalization Pathways on YouTube," arXiv:1908.08313 (2019), https://arxiv.org/abs/1908.08313.

31. Evan Osnos, "Can Mark Zuckerberg Fix Facebook before It Breaks Democracy?" *New Yorker,* September 10, 2018, https://www.newyorker.com/magazine/2018/09/17/can-mark-zuckerberg-fix-facebook-before-it-breaks-democracy.

32. James Bridle, "Something Is Wrong on the Internet," *Medium,* https://medium.com/@jamesbridle/something-is-wrong-on-the-internet-c39c471271d2.

33. James Bridle, *New Dark Age* (New York: Verso, 2018), 230.

34. Max Fisher and Amanda Taub, "On YouTube's Digital Playground, an Open Gate for Pedophiles," New York Times, June 3, 2019, at https://www.nytimes.com/2019/06/03/world/americas/youtube-pedophiles.html.

35. Frank Pasquale, "Reclaiming Egalitarianism in the Political Theory of Campaign Finance Reform," *University of Illinois Law Review* (2008): 599–660.

36. Rebecca Hersher, "What Happened When Dylann Roof Asked Google for Information about Race?" *Houston Public Media,* January 10, 2017, http://www.houstonpublicmedia.org/npr/2017/01/10/508363607/what-happened-when-dylann-roof-asked-google-for-information-about-race/.

37. Hiroko Tabuchi, "How Climate Change Deniers Rise to the Top in Google Searches," *New York Times,* December 9, 2017, https://www.nytimes.com/2017/12/29/climate/google-search-climate-change.html.

38. Noble, *Algorithms of Oppression.*

39. Cadwalladr, "Google, Democracy and the Truth."

40. Ben Guarino, "Google Faulted for Racial Bias in Image Search Results for Black Teenagers," *Washington Post,* June 10, 2016, https://www.washingtonpost.com/news/morning-mix/wp/2016/06/10/google-faulted-for-racial-bias-in-image-search-results-for-black-teenagers/?utm_term=.1a3595bb8624. Also Tom Simonite, "When It Comes to Gorillas, Google Photos Remains Blind," *Wired,* January 11, 2018, https://www.wired.com/story/when-it-comes-to-gorillas-google-photos-remains-blind/. On the larger problem, see Aylin Caliskan, Joanna J. Bryson, and Arvind

Nararayan, "Semantics Derived Automatically from Language Corpora Contain Human-Like Biases," *Science* 356 (2017): 183–186.

41. Annalee Newitz, "Opinion: A Better Internet Is Waiting for Us," *New York Times*, November 30, 2019, https://www.nytimes.com/interactive/2019/11/30/opinion/social-media-future.html.

42. Newitz, "Opinion: A Better Internet is Waiting for Us."

43. "Preamble," The Syllabus, accessed May 15, 2020, https://the-syllabus.com/preamble-mission/; Maurits Martijn, "The Most Important Technology Critic in the World Was Tired of Knowledge Based on Clicks. So He Built an Antidote," *Correspondent*, March 26, 2020, https://thecorrespondent.com/369/the-most-important-technology-critic-in-the-world-was-tired-of-knowledge-based-on-clicks-so-he-built-an-antidote/789698745-92d7c0ee.

44. I borrow the term "infodemic" from Kate Starbird. Kate Starbird, "How to Cope with an Infodemic," Brookings TechStream, April 27, 2020, https://www.brookings.edu/techstream/how-to-cope-with-an-infodemic/.

45. Tim Wu, The Curse of Bigness: Antitrust in the New Gilded Age (New York: Columbia Global Reports, 2019); Matt Stoller, Sarah Miller, and Zephyr Teachout, Addressing Facebook and Google's Harms Through a Regulated Competition Approach, White Paper of the American Economic Liberties Project (2020).

46. Sanjukta Paul, "Antitrust as Allocator of Coordination Rights," *University of California at Los Angeles Law Review* 67 (forthcoming 2020).

47. Cat Ferguson, "Searching for Help," *Verge*, September 7, 2017, https://www.theverge.com/2017/9/7/16257412/rehabs-near-me-google-search-scam-florida-treatment-centers.

48. Andrew McAfee and Erik Brynjolfsson, *Machine, Platform, Crowd: Harnessing our Digital Future* (New York: Norton 2017).

49. David Dayen, "Google Is So Big, It Is Now Shaping Policy to Combat the Opioid Epidemic: And It's Screwing It Up," *Intercept*, October 17, 2017, https://theintercept.com/2017/10/17/google-search-drug-use-opioid-epidemic/.

50. Ferguson, "Searching for Help"; Dayen, "Google Is So Big."

51. 2017 Fla. Laws 173 (codified at Fla. Stat. §§ 397.55, 501.605, 501.606, 817.0345).

52. Ryan Singel, "Feds Pop Google for $500M for Canadian Pill Ads," *Wired*, August 24, 2011, https://www.wired.com/2011/08/google-drug-fine/.

53. Drew Harwell, "AI Will Solve Facebook's Most Vexing Problems, Mark Zuckerberg Says. Just Don't Ask When or How," *Washington Post,* April 11, 2018, https://www.washingtonpost.com/news/the-switch/wp /2018/04/11/ai-will-solve-facebooks-most-vexing-problems-mark -zuckerberg-says-just-dont-ask-when-or-how/?utm_term=.a0a7c340ac66.

54. Instagram, "Community Guidelines," https://help.instagram.com /477434105621119; also Sarah Jeong, "I Tried Leaving Facebook. I Couldn't," *Verge,* April 28, 2018, https://www.theverge.com/2018/4/28 /17293056/facebook-deletefacebook-social-network-monopoly.

55. Joel Ross, Lilly Irani, M. Six Silberman, Andrew Zaldivar, and Bill Tomlinson, "Who Are the Crowdworkers? Shifting Demographics in Mechanical Turk," *CHI EA '10: ACM Conference on Human Factors in Computing Systems* (2010), 2863–2872.

56. Jaron Lanier, *You Are Not a Gadget: A Manifesto* (New York: Knopf, 2010).

57. Julia Powles, "The Case That Won't Be Forgotten," *Loyola University Chicago Law Journal* 47 (2015): 583–615; Frank Pasquale, "Reforming the Law of Reputation," *Loyola University Chicago Law Journal* 47 (2015): 515–539.

58. Michael Hiltzik, "President Obama Schools Silicon Valley CEOs on Why Government Is Not Like Business," *Los Angeles Times,* October 17, 2016, https://www.latimes.com/business/hiltzik/la-fi-hiltzik-obama-silicon -valley-20161017-snap-story.html.

59. Case C-131 / 12, Google Spain SL v. Agencia Española de Protección de Datos (AEPD), (May 13, 2014), http://curia.europa.eu/juris/document /document.jsf?text=&docid=152065&doclang=EN.

60. The story has not been disappeared entirely; it may well appear prominently when the husband's name is searched or as a result in response to other queries. The privacy claim here is limited to searches of the widow's name.

61. Julia Powles, "Results May Vary: Border Disputes on the Frontlines of the 'Right to be Forgotten,'" *Slate,* February 25, 2015, http://www.slate.com /articles/technology/future_tense/2015/02/google_and_the_right_to_be _forgotten_should_delisting_be_global_or_local.html.

62. Justin McCurry, "Japanese Court Rules against Paedophile in 'Right to be Forgotten' Online Case," *Guardian,* February 1, 2017, https://www

.theguardian.com/world/2017/feb/02/right-to-be-forgotten-online-suffers
-setback-after-japan-court-ruling.

63. For more general appreciation of human flexibility, see Geoff Colvin, *Humans Are Underrated: What High Achievers Know That Brilliant Machines Never Will* (New York: Penguin, 2015); Hubert L. Dreyfus and Stuart E. Dreyfus, *Mind over Machine: The Power of Human Expertise and Intuition in the Era of the Computer* (New York: Free Press, 1988).

64. Dan Solove, *Future of Reputation: Gossip, Rumor and Privacy on the Internet* (New Haven, CT: Yale University Press, 2007); Sarah Esther Lageson, *Digital Punishment Privacy, Stigma, and the Harms of Data-Driven Criminal Justice* (Oxford: Oxford University Press, 2020).

65. Sarah T. Roberts, *Behind the Screen: Content Moderation in the Shadows of Social Media* (New Haven: Yale University Press, 2019); Lilly Irani, "Justice for 'Data Janitors,'" *Public Books,* January 15, 2015, http://www
.publicbooks.org/nonfiction/justice-for-data-janitors.

66. Phillips, *The Oxygen of Amplification;* danah boyd, "Hacking the Attention Economy," *Points: Data & Society* (blog), January 5, 2017, https://points
.datasociety.net/hacking-the-attention-economy-9fa1daca7a37; see also Pasquale, "Platform Neutrality"; Frank Pasquale, "Asterisk Revisited: Debating a Right of Reply on Search Results," *Journal of Business and Technology Law* 3 (2008): 61–86.

67. Mark Bergen, "YouTube Executives Ignored Warnings, Letting Toxic Videos Run Rampant," *Bloomberg News,* April 2, 2019, https://www
.bloomberg.com/news/features/2019-04-02/youtube-executives-ignored
-warnings-letting-toxic-videos-run-rampant.

68. My position here echoes the name of the anti–*Citizens United* group Free Speech for People, as I believe the doctrine of "computer speech" could evolve in the same troubling ways that corporate speech doctrine has.

69. James Grimmelmann, "Copyright for Literate Robots," *Iowa Law Review* 101 (2016): 657–682.

70. Danielle K. Citron and Robert Chesney, "Deep Fakes: A Looming Challenge for Privacy, Democracy, and National Security," *California Law Review* 107 (2019): 1753–1819.

71. Mary Anne Franks and Ari Ezra Waldman, "Sex, Lies, and Videotape: Deep Fakes and Free Speech Delusions," *Maryland Law Review* 78 (2019): 892–898. Whatever autonomy interests may be asserted by the creators of

such deep fakes, they must be weighed against the audience's interest in an undistorted public sphere. Robyn Caplan, Lauren Hanson, and Joan Donovan, *Dead Reckoning: Navigating Content Moderation after Fake News* (New York: Data & Society, 2018), https://datasociety.net/library /dead-reckoning/; Alice Marwick and Rebecca Lewis, *Media Manipulation and Disinformation Online* (New York: Data & Society, 2015), https://datasociety.net/pubs/oh/DataAndSociety_MediaManipulationAn dDisinformationOnline.pdf.

72. James Grimmelmann, "Listeners' Choices," *University of Colorado Law Review* 90 (2018): 365–410.

73. The BCAP Code: The UK Code of Broadcast Advertising, Appendix 2 (September 1, 2010), http://www.asa.org.uk/uploads/assets/uploaded /e6e8b10a-20e6-4674-a7aa6dc15aa4f814.pdf. The US Federal Communications Commission has twice considered the issue but done nothing. See also Ryan Calo, "Digital Market Manipulation," *George Washington Law Review* 82 (2014): 995–1051.

74. Bolstering Online Transparency Act, California Business & Professions Code §§ 17940–43 (2019).

75. An Act to Add Article 10 (commencing with Section 17610) to Chapter 1 of Part 3 of Division 7 of the Business and Professions Code, relating to advertising, A.B. 1950, 2017–2018 Reg. Sess. (January 29, 2018), http:// leginfo.legislature.ca.gov/faces/billNavClient.xhtml?bill_id =201720180AB1950.

76. Frederic Lardinois, "The Google Assistant Will Soon Be Able to Call Restaurants and Make a Reservation for You," TechCrunch, May 8, 2018, https://techcrunch.com/2018/05/08/the-google-assistant-will-soon-be-able -to-call-restaurants-and-make-a-reservations-for-you/ (see video at 0:24).

77. Yaniv Leviathan, "Google Duplex: An AI System for Accomplishing Real-World Tasks over the Phone," Google AI Blog, May 8, 2018, https://ai .googleblog.com/2018/05/duplex-ai-system-for-natural-conversation.html ?fbclid=IwAR0K2LGSd9wLPeN3lejGQhj8TAaJifQL0aKPZ9eivpBoQgTA wakOTbPPIMw.

78. Alice Gregory, "R U There? A New Counselling Service Harnesses the Power of the Text Message," *New Yorker*, February 2, 2015, https://www .newyorker.com/magazine/2015/02/09/r-u.

79. Nick Statt, "Google Now Says Controversial AI Voice Calling System Will Identify Itself to Humans," *Verge*, May 10, 2018, https://www.theverge

.com/2018/5/10/17342414/google-duplex-ai-assistant-voice-calling-identify -itself-update.

80. Masahiro Mori, "The Uncanny Valley: The Original Essay by Masahiro Mori," trans. Karl F. MacDorman and Norri Kageki, *IEEE Spectrum*, June 12, 2012, https://spectrum.ieee.org/automaton/robotics/humanoids /the-uncanny-valley.

81. Natasha Lomas, "Duplex Shows Google Failing at Ethical and Creative AI Design," TechCrunch, May 10, 2018, https://techcrunch.com/2018/05/10 /duplex-shows-google-failing-at-ethical-and-creative-ai-design/.

82. James Vincent, "Google's AI Sounds Like a Human on the Phone— Should We Be Worried?" *Verge*, May 9, 2018, https://www.theverge.com /2018/5/9/17334658/google-ai-phone-call-assistant-duplex-ethical-social -implications.

83. Rachel Metz, "Facebook's Top AI Scientist Says It's 'Dust' without Artificial Intelligence," *CNN*, December 5, 2018, https://www.cnn.com/2018/12/05 /tech/ai-facebook-lecun/index.html.

84. The AI in the automated public sphere is an example of what Introna and Wood call *silent,* as opposed to *salient,* technology. It is embedded, obscure, flexible, and mobile (applied via software), as opposed to more noticeably surface, transparent, and fixed. Lucas Introna and David Wood, "Picturing Algorithmic Surveillance: The Politics of Facial Recognition Systems," *Surveillance and Society* 2, no. 2/3 (2002):177–198.

85. Sam Thielman, "Q&A: Mike Ananny, On Facebook, Facts, and a Marketplace of Ideas," *Columbia Journalism Review,* April 5, 2018, https://www.cjr.org/tow_center/qa-usc-annenbergs-mike-ananny-on -facebooks-fact-checkers-and-the-problem-with-a-marketplace-of-ideas .php; Mike Ananny, "The Partnership Press: Lessons for Platform-Publisher Collaborations as Facebook and News Outlet Teams to Fight Misinformation," Tow Center for Digital Journalism Report, April 4, 2018, https://www.cjr.org/tow_center_reports/partnership-press -facebook-news-outlets-team-fight-misinformation.php.

86. Tim O'Reilly, "Media in the Age of Algorithms," *Medium,* November 11, 2016, https://medium.com/the-wtf-economy/media-in-the-age-of -algorithms-63e80b9b0a73#.9l86jw9r4.

87. Abby Ohlheiser, "Three Days after Removing Human Editors, Facebook Is Already Trending Fake News," *Washington Post*, August 29, 2016, https://www.washingtonpost.com/news/the-intersect/wp/2016/08/29/a-fake

-headline-about-megyn-kelly-was-trending-on-facebook/?utm_term=
.f857ac42b2e9; "Facebook: 'No Evidence' Conservative Stories Were
Suppressed," *CBS News,* May 10, 2016, http://www.cbsnews.com/news
/facebook-no-evidence-conservative-stories-trending-suppressed-gizmodo/.

88.   Josh Sternberg, "Layoffs and Local Journalism," *Media Nut,* May 14, 2020,
at https://medianut.substack.com/p/layoffs-and-local-journalism.

89.   Rachael Revesz, "Steve Bannon's Data Firm in Talks for Lucrative White
House Contracts," *Independent,* November 23, 2016, http://www
.independent.co.uk/news/world/americas/cambridge-analytica-steve
-bannon-robert-rebekah-mercer-donald-trump-conflicts-of-interest
-white-a7435536.html; Josh Feldman, "CIA Concluded Russia Intervened
in Election to Help Trump, WaPo Reports," *Mediaite,* December 9, 2016,
http://www.mediaite.com/online/cia-concluded-russia-intervened-in
-election-to-help-trump-wapo-reports/.

90.   Will Oremus, "The Prose of the Machines," *Slate,* July 14, 2014, http://
www.slate.com/articles/technology/technology/2014/07/automated
_insights_to_write_ap_earnings_reports_why_robots_can_t_take
_journalists.html.

91.   Thorstein Veblen, Absentee Ownership and Business Enterprise in Recent
Times (London: George Allen & Unwin, 1923); Christopher Meek,
Warner Woodworth, and W. Gibb Dyer, *Managing by the Numbers:
Absentee Ownership and the Decline of American Industry* (New York:
Addison-Wesley, 1988).

92.   The problem is even worse overseas. Mark Zuckerberg once remarked, "A
squirrel dying in your front yard may be more relevant to your interests
right now than people dying in Africa." Jaigris Hodson, "When a Squirrel
Dies: The Rapid Decline of Local News," The Conversation, September 13,
2017, at https://theconversation.com/when-a-squirrel-dies-the-rapid
-decline-of-local-news-82120. While callous, the statement was an honest
reflection of the neo-colonial mentality of massive technology firms.
Nick Couldry and Ulises A. Mejias, *The Costs of Connection How Data
Is Colonizing Human Life and Appropriating It for Capitalism* (Redwood
City: Stanford University Press, 2019).

93.   Rahel Jaeggi, *Alienation* (trans. Frederick Neuhouser and Alan E. Smith)
(New York: Columbia University Press, 2014), 1.

94.   Britt Paris and Joan Donovan, *Deepfakes and Cheap Fakes: The Manipulation of Audio and Visual Evidence* (New York: Data & Society, 2019).

5. MACHINES JUDGING HUMANS

1. Mike Butcher, "The Robot-Recruiter Is Coming—VCV's AI Will Read Your Face in a Job Interview," *TechCrunch,* April 23, 2019, https:// techcrunch.com/2019/04/23/the-robot-recruiter-is-coming-vcvs-ai-will -read-your-face-in-a-job-interview/.

2. Miranda Bogen and Aaron Rieke, *Help Wanted: An Examination of Hiring Algorithms, Equity, and Bias* (Washington, DC: Upturn, 2018), https://www.upturn.org/static/reports/2018/hiring-algorithms/files /Upturn%20—%20Help%20Wanted%20-%20An%20Exploration%20 of%20Hiring%20Algorithms,%20Equity%20and%20Bias.pdf.

3. There are numerous examples of person-judging technology reinscribing and reinforcing old forms of unearned privilege and unfair disadvantage. Ruha Benjamin, *Race after Technology: Abolitionist Tools for the New Jim Code* (Cambridge, UK: Polity Press, 2019); Safiya Umoja Noble, *Algorithms of Oppression: How Search Engines Reinforce Racism* (New York: New York University Press, 2018).

4. "Fairness, Accountability, and Transparency in Machine Learning," FAT / ML, accessed May 1, 2020, https://www.fatml.org/.

5. James Vincent, "All of These Faces Are Fake Celebrities Spawned by AI," *Verge,* October 30, 2017, https://www.theverge.com/2017/10/30/16569402 /ai-generate-fake-faces-celebs-nvidia-gan

6. Daniel J. Solove, "Privacy and Power: Computer Databases and Metaphors for Information Privacy," *Stanford Law Review* 53 (2001): 1393–1462; Philip K. Dick, *Ubik* (New York: Doubleday, 1969).

7. Jathan Sadowski and Frank Pasquale, "The Spectrum of Control: A Social Theory of the Smart City," *First Monday* 20, no. 7 (2015), https://doi.org /10.5210/fm.v20i7.5903.

8. For a discussion of computer scientists using inefficiency in design to "make space for human values," see Paul Ohm and Jonathan Frankel, "Desirable Inefficiency," *Florida Law Review* 70, no. 4 (2018): 777–838. For a discussion of how design inefficiency contradicts this urge of coders, see Clive Thompson, "Coders' Primal Urge to Kill Inefficiency— Everywhere," *Wired,* March 19, 2019, https://www.wired.com/story /coders-efficiency-is-beautiful/?curator=TechREDEF.

9. Arjun Jayadev and Samuel Bowles, "Guard Labor," *Journal of Development Economics* 79, no. 2 (2006): 328–348.

10. I. Bennett Capers, "Afrofuturism, Critical Race Theory, and Policing in the Year 2044," *New York University Law Review* 94, no. 1 (2019): 1–60. See also William H. Simon, "In Defense of the Panopticon," *Boston Review* 39, no. 5 (2014): 58–74.

11. Anthony Funnell, "Internet of Incarceration: How AI Could Put an End to Prisons As We Know Them," *ABC News*, August 14, 2017, https://www .abc.net.au/news/2017-08-14/how-ai-could-put-an-end-to-prisons-as-we -know-them/8794910.

12. Ibid.

13. Benjamin, *Race after Technology*, 137.

14. Baz Dreisinger, *Incarceration Nations: A Journey to Justice in Prisons around the World* (New York: Other Press, 2016).

15. Big Brother Watch, "Face Off: Stop the Police Using Authoritarian Facial Recognition Cameras," March 15, 2019, https://bigbrotherwatch.org.uk /all-campaigns/face-off-campaign/#breakdown.

16. Kevin Arthur, "Hartzog and Selinger: Ban Facial Recognition," *Question Technology* (blog), August 4, 2018, https://www.questiontechnology.org /2018/08/04/hartzog-and-selinger-ban-facial-recognition/; Woodrow Hartzog and Evan Selinger, "Surveillance As Loss of Obscurity," *Washington and Lee Law Review* 72 (2015): 1343–1388.

17. Joy Buolamwini and Timnit Gebru, "Gender Shades: Intersectional Accuracy Disparities in Commercial Gender Classification," *Proceedings of Machine Learning Research* 81 (2018): 1–15, http://proceedings.mlr .press/v81/buolamwini18a/buolamwini18a.pdf.

18. Richard Feloni, "An MIT Researcher Who Analyzed Facial Recognition Software Found Eliminating Bias in AI Is a Matter of Priorities," *Business Insider,* January 23, 2019, https://www.businessinsider.com/biases-ethics -facial-recognition-ai-mit-joy-buolamwini-2019-1.

19. Jacob Snow, "Amazon's Face Recognition Falsely Matched 28 Members of Congress with Mugshots," ACLU, July 26, 2018, https://www.aclu.org /blog/privacy-technology/surveillance-technologies/amazons-face -recognition-falsely-matched-28.

20. Zoé Samudzi, "Bots Are Terrible at Recognizing Black Faces. Let's Keep It that Way," *Daily Beast,* February 8, 2019, https://www.thedailybeast.com /bots-are-terrible-at-recognizing-black-faces-lets-keep-it-that-way.

21. Alondra Nelson (@alondra), "Algorithmic Accountability Is Tremendously Important. Full Stop," Twitter, February 9, 208, 5:48 p.m., https://twitter.com/alondra/status/962095979553009665.

22. Evan Selinger and Woodrow Hartzog, "What Happens When Employers Can Read Your Facial Expressions?," New York Times, October 17, 2019, https://www.nytimes.com/2019/10/17/opinion/facial-recognition-ban .html.

23. For the full implications of such universalizing surveillance, see John Gillion and Torin Monahan, *Supervision: An Introduction to the Surveillance Society* (Chicago: University of Chicago Press, 2012); and Ian G. R. Shaw, *Predator Empire: Drone Warfare and Full Spectrum Dominance* (Minneapolis: University of Minnesota Press, 2018).

24. Evan Selinger and Woodrow Hartzog, "Opinion: It's Time for an About-Face on Facial Recognition," *Christian Science Monitor,* June 22, 2015, https://www.csmonitor.com/World/Passcode/Passcode-Voices/2015 /0622/Opinion-It-s-time-for-an-about-face-on-facial-recognition.

25. Amy Hawkins, "Beijing's Big Brother Tech Needs African Faces," *Foreign Policy,* July 24, 2018, https://foreignpolicy.com/2018/07/24/beijings-big -brother-tech-needs-african-faces/.

26. Darren Byler, "China's Hi-Tech War on Its Muslim Minority," *Guardian,* April 11, 2019, https://www.theguardian.com/news/2019/apr/11/china -hi-tech-war-on-muslim-minority-xinjiang-uighurs-surveillance-face -recognition; Chris Buckley, Paul Mozur, and Austin Ramzy, "How China Turned a City into a Prison," *New York Times,* April 4, 2019, https://www.nytimes.com/interactive/2019/04/04/world/asia/xinjiang -china-surveillance-prison.html; Mozur, "One Month, 500,000 Face Scans."

27. Luke Stark, "Facial Recognition Is the Plutonium of AI," *XRDS: Crossroads, The ACM Magazine for Students* 25, no. 3 (Spring 2019): 50–55; Woodrow Hartzog and Evan Selinger, "Facial Recognition Is the Perfect Tool for Oppression," *Medium,* August 2, 2018, https://medium.com/s /story/facial-recognition-is-the-perfect-tool-for-oppression-bc2a08f0fe66.

28. Robert Pear, "On Disability and on Facebook? Uncle Sam Wants to Watch What You Post," *New York Times,* March 10, 2019, https://www .nytimes.com/2019/03/10/us/politics/social-security-disability-trump -facebook.html.

29. Julie E. Cohen, *Configuring the Networked Self: Law, Code, and the Play of Everyday Practice* (New Haven: Yale University Press, 2012). Cohen's conception of semantic discontinuity also applies online, and it defends spaces of anonymity there as well.

30. Margaret Hu, "Big Data Blacklisting," *Florida Law Review* 67 (2016): 1735-1809; Eric J. Mitnick, "Procedural Due Process and Reputational Harm: Liberty as Self-Invention," *University of California Davis Law Review* 43 (2009): 79–142.

31. Sam Levin, "Face-Reading AI Will Be Able to Predict Your Politics and IQ, Professor Says," *Guardian,* September 12, 2017, https://www.theguardian.com/technology/2017/sep/12/artificial-intelligence-face-recognition-michal-kosinski.

32. Sam Biddle, "Troubling Study Says Artificial Intelligence Can Predict Who Will Be Criminals Based on Facial Features," *Intercept,* November 18, 2016, https://theintercept.com/2016/11/18/troubling-study-says-artificial-intelligence-can-predict-who-will-be-criminals-based-on-facial-features/.

33. Matt McFarland, "Terrorist or Pedophile? This Start-Up Says It Can Out Secrets by Analyzing Faces," *Washington Post,* May 24, 2016, https://www.washingtonpost.com/news/innovations/wp/2016/05/24/terrorist-or-pedophile-this-start-up-says-it-can-out-secrets-by-analyzing-faces/?noredirect=on&utm_term=.5a060615a547. Faception also presumes to classify persons in more positive ways. Faception, "Our Technology," https://www.faception.com/our-technology.

34. Dan McQuillan, "People's Councils for Ethical Machine Learning," *Social Media + Society,* April–June 2018: 1–10.

35. Judea Pearl and Dana McKenzie, *The Book of Why: The New Science of Cause and Effect* (New York: Basic Books, 2018).

36. Frank Pasquale and Glyn Cashwell, "Prediction, Persuasion, and the Jurisprudence of Behaviourism," *University of Toronto Law Journal* 68 supp. (2018): 63–81. We should also be cautious of data based on practices that have been demonstrated to be discriminatory. Rashida Richardson, Jason M. Schultz, and Kate Crawford, "Dirty Data, Bad Predictions: How Civil Rights Violations Impact Police Data, Predictive Policing Systems, and Justice," *New York University Law Review* 94 (2019): 192–233.

37. Fabio Ciucci, "AI (Deep Learning) Explained Simply," *Data Science Central* (blog), November 20, 2018, https://www.datasciencecentral.com

/profiles/blogs/ai-deep-learning-explained-simply ("MLs fail when the older data gets less relevant or wrong very soon and often. The task or rules learned must keep the same, or at most rarely updated, so you can re-train.").

38. Kiel Brennan-Marquez, "Plausible Cause: Explanatory Standards in the Age of Powerful Machines," *Vanderbilt Law Review* 70 (2017): 1249–1302.

39. Robert H. Sloan and Richard Wagner, "Avoiding Alien Intelligence: A Plea for Caution in the Use of Predictive Systems," https://papers.ssrn .com/sol3/papers.cfm?abstract_id=3163664; Guido Noto La Diega, "Against the Dehumanization of Decision-Making: Algorithmic Decisions at the Crossroads of Intellectual Property, Data Protection, and Freedom of Information," *Journal of Intellectual Property, Information Technology and E-commerce Law* 9, no. 1 (2018): https://www.jipitec.eu/issues/jipitec -9-1-2018/4677.

40. Patrick Tucker, "The US Military Is Creating the Future of Employee Monitoring," *Defense One*, March 26, 2019, https://www.defenseone.com /technology/2019/03/us-military-creating-future-employee-monitoring /155824/.

41. On the polygraph's unreliability and its banishment from employment contexts in the US, see Joseph Stromberg, "Lie Detectors: Why They Don't Work, and Why Police Use Them Anyway," Vox, Dec 15, 2014, at https://www.vox.com/2014/8/14/5999119/polygraphs-lie-detectors-do -they-work.

42. This quote is from her interview in Martin Ford, *Architects of Intelligence: The Truth about AI from the People Building It* (Birmingham, UK: Packt Publishing, 2018), 217.

43. Ifeoma Ajunwa, Kate Crawford, and Jason Schulz, "Limitless Worker Surveillance," *California Law Review* 105 (2017): 735–776.

44. Marika Cifor, Patricia Garcia, T. L. Cowan, Jasmine Rault, Tonia Sutherland, Anita Say Chan, Jennifer Rode Anna Lauren Hoffmann, Niloufar Salehi, and Lisa Nakamura, *Feminist Data Manifest-No*, https://www.manifestno.com/.

45. Danielle Keats Citron and Frank Pasquale, "The Scored Society: Due Process for Automated Predictions," *Washington Law Review* 89 (2014): 1–34. One of the most ambitious proposals to address the many threats to privacy discussed in this chapter is U.S. Senator Sherrod Brown's Data Accountability and Transparency Act of 2020.

46.   Privacy International, "Case Study: Fintech and the Financial Exploitation of Customer Data," August 30, 2017, https://www.privacyinternational .org/case-studies/757/case-study-fintech-and-financial-exploitation -customer-data.

47.   Ibid.

48.   Rachel Botsman, "Big Data Meets Big Brother as China Moves to Rate its Citizens," *Wired*, Oct. 21, 2017. For a scholarly analysis of SCS's, see Yu-Jie Chen, Ching-Fu Lin, and Han-Wei Liu, "'Rule of Trust': The Power and Perils of China's Social Credit Megaproject," *Columbia Journal of Asian Law* 32 (2018): 1–36.

49.   James Rufus Koren, "Some Lenders Are Judging You on Much More Than Your Finances," *Los Angeles Times,* December 19, 2015, https://www .latimes.com/business/la-fi-new-credit-score-20151220-story.html.

50.   Ryen W. White, P. Murali Doraiswamy, and Eric Horvitz, "Detecting Neurogenerative Disorders from Web Search Signals," *NPJ Digital Medicine* 1 (2018): article no. 8, https://www.nature.com/articles/s41746 -018-0016-6.pdf.

51.   Louise Seamster and Raphaël Charron-Chénier, "Predatory Inclusion and Education Debt: Rethinking the Racial Wealth Gap," *Social Currents* 4 (2017): 199–207.

52.   Donald Bradley, "KC Man Pays $50,000 Interest on $2,500 in Payday Loans," *Kansas City Star,* May 17, 2016, https://www.kansascity.com/news /local/article78174997.html.

53.   "I Borrowed £150 . . . Now I Owe £10k: Christmas Warning over Payday Loans," *Sun,* December 16, 2012, updated April 4, 2016, https://www .thesun.co.uk/archives/news/342926/i-borrowed-150-now-i-owe-10k/.

54.   Kevin P. Donovan and Emma Park, "Perpetual Debt in the Silicon Savannah," *Boston Review,* September 20, 2019, at http://bostonreview.net /class-inequality-global-justice/kevin-p-donovan-emma-park-perpetual -debt-silicon-savannah.

55.   Tressie McMillan Cottom, *Lower Ed: The Troubling Rise of For-Profit Colleges in the New Economy* (New York: The New Press, 2018).

56.   Lauren Berlant, *Cruel Optimism* (Durham, NC: Duke University Press, 2011); David J. Blacker, *The Falling Rate of Earning and the Neoliberal Endgame* (Washington, DC: Zero Books, 2013); Andrew McGettigan, *The Great University Gamble: Money, Markets and the Future of Higher*

Education (London: Pluto Press, 2013); Christopher Newfield, *The Great Mistake: How We Wrecked Public Universities and How We Can Fix Them* (Baltimore: Johns Hopkins University Press, 2016).

57. Frank Pasquale, "Democratizing Higher Education: Defending and Extending Income Based Repayment Programs," *Loyola Consumer Law Review* 28 (2015): 1–30.

58. The United States has such a program as well, but it is so poorly run that it is hard to recommend it. For example, it supposedly forgives loans after ten years in public service, but of the first group of thousands of claimants, less than 1 percent were approved. Annie Nova, "Here Are the Facts about Public Service Loan Forgiveness," CNBC, December 21, 2018, at https://www.cnbc.com/2018/12/21/1-percent-of-people-were-approved -for-public-service-loan-forgiveness.html. The private servicers behind US student loans earn more money the longer they can keep borrowers on the hook for repayments, so they are notoriously bad at processing paperwork (frequently losing forms). The clear lesson of the US experience is that a simple and clear program, not run by entities with an interest in perpetuating debt, is best.

59. Scholars are also recognizing the limits of loans as a form of social provision. Abbye Atkinson, "Rethinking Credit as Social Provision," *Stanford Law Review* 71 (2019): 1093–1162.

60. Adam Kotsko, *Creepiness* (Washington, DC: Zero Books, 2015); see also Omer Tene and Jules Polonetsky, "A Theory of Creepy: Technology, Privacy, and Shifting Social Norms," *Yale Journal of Law and Technology* 16 (2014): 59–102, for examples in the context of privacy where "creepy behavior pushes against traditional social norms."

61. For an example, see Jessica Leber, "This New Kind of Credit Score Is All Based on How You Use Your Cell Phone," *Fast Company,* April 27, 2016, https://www.fastcompany.com/3058725/this-new-kind-of-credit -score-is-all-based-on-how-you-use-your-cellphone, which discusses the assurances of Equifax that their privacy and data security practices were appropriate.

62. ACLU of California, "Metadata: Piecing Together a Privacy Solution," February 2014, https://www.aclunc.org/sites/default/files/Metadata%20 report%20FINAL%202%2021%2014%20cover%20%2B%20inside%20 for%20web%20%283%29.pdf.

63. Ajunwa, Crawford, and Schultz, "Limitless Workplace Surveillance."

64. Shoshana Zuboff, *The Age of Surveillance Capitalism: The Fight for a Human Future at the New Frontier of Power* (New York: PublicAffairs, 2019).

65. Davis Polk, "Time to Get Serious about Microchipping Employees and Biometric Privacy Laws," *Law Fuel,* February 14, 2019, http://www .lawfuel.com/blog/time-to-get-serious-about-microchipping-employees -and-biometric-privacy-laws/.

66. Clement Yongxi Chen and Anne S. Y. Cheung, "The Transparent Self under Big Data Profiling: Privacy and Chinese Legislation on the Social Credit System," *Journal of Comparative Law* 12 (2017): 356–378.

67. State Council Notice, *Planning Outline for the Construction of a Social Credit System,* original text and English translation available at *China Copyright and Media* (June 14, 2014, updated April 25, 2015), https:// chinacopyrightandmedia.wordpress.com/2014/06/14/planning-outline -for-the-construction-of-a-social-credit-system-2014-2020/. The official, original text is available through the Chinese government website, http://www.gov.cn/zhengce/content/2014-06/27/content_8913.htm.

68. Evelyn Cheng and Shirley Tay, "China Wants to Track and Grade Each Citizen's Actions—It's in the Testing Phase," CNBC, July 25, 2019, https:// www.cnbc.com/2019/07/26/china-social-credit-system-still-in-testing -phase-amid-trials.html. See also Sophia Yan, "The Village Testing China's Social Credit System: Driven by Big Data, Its Residents Earn a Star Rating," *South China Morning Post,* June 2, 2019, https://www.scmp .com/magazines/post-magazine/long-reads/article/3012574/village -testing-chinas-social-credit-system.

69. Cheng and Tay, "China Wants to Track and Grade Each Citizen's Actions." Cheng and Tay cite the Credit China report on testing of a social credit system in Rongcheng, Shandong, released February 5, 2018 and available through the Chinese government website, https://www .creditchina.gov.cn/chengxinwenhua/chengshichengxinwenhua/201802 /t20180205_108168.html.

70. Cheng and Tay, "China Wants to Track and Grade Each Citizen's Actions."

71. Chen and Cheung coined the term "rippling" for this effect. Yongxi Chen and Anne S. Y. Cheung, "The Transparent Self under Big Data Profiling: Privacy and Chinese Legislation on the Social Credit System," *Journal of Comparative Law* 12 (2017): 356–378. It is reminiscent of the "collateral

consequences" of criminal conviction or debt default in the United States. See, for example, Michael Pinard, "Collateral Consequences of Criminal Convictions: Confronting Issues of Race and Dignity," *New York University Law Review* 85 (2010): 457–534.

72. *Southern Metropolis Daily*, "The Parents Owed Their Debts, and 31 Children Were Recommended by the Court to Drop Out of School," *21Jingji*, July 18, 2018, at https://m.21jingji.com/article/20180718/herald/b a0ff13df6dcb5196cdcb6f886e4726b.html. This aspect of reputational assessment is described and critiqued in Hu Jianmiao, Can the Children of Those Who Run Red Lights be Restricted from Attending School, Legal Daily, January 13, 2016, at http://opinion.people.com.cn/n1/2016 /0113/c1003-28046944.html. Hu's critique suggests robust bases for challenging networked consequences within China's legal system.

73. Jay Stanley, "China's Nightmarish Citizen Scores Are a Warning for Americans," ACLU (blog), October 5, 2015, https://www.aclu.org/blog /privacy-technology/consumer-privacy/chinas-nightmarish-citizen -scores-are-warning-americans.

74. Perhaps anticipating negative reputational effects, Hong Kong protesters against the Extradition Bill (2019) have used masks, cash Octopus cards, and lasers to disrupt systems of facial recognition that might perma- nently stigmatize them as dissidents or "untrustworthy."

75. Gilles Deleuze, "Postscript on the Societies of Control," *October* 59 (1992): 3–7; Frank Pasquale, "The Algorithmic Self," *Hedgehog Review*, Spring 2015.

76. Nathan Vanderklippe, "Chinese Blacklist an Early Glimpse of Sweeping New Social-Credit Control," *Globe and Mail,* January 3, 2018, updated January 18, 2018, https://www.theglobeandmail.com/news/world/chinese -blacklist-an-early-glimpse-of-sweeping-new-social-credit-control /article37493300/; Frank Pasquale and Danielle Keats Citron, "Promoting Innovation while Preventing Discrimination: Policy Goals for the Scored Society," *Washington Law Review* 89 (2014): 1413–1424; Danielle Keats Citron and Frank Pasquale, "The Scored Society."

77. Dan McQuillan, "Algorithmic States of Exception," *European Journal of Cultural Studies* 18 (2015): 564–576.

78. Jathan Sadowski and Frank Pasquale, "The Spectrum of Control: A Social Theory of the Smart City," *First Monday* 20, no. 7 (July 6, 2015), https:// firstmonday.org/article/view/5903/4660.

79. Violet Blue, "Your Online Activity Is Now Effectively a Social 'Credit Score,'" https://www.engadget.com/2020/01/17/your-online-activity -effectively-social-credit-score-airbnb/.

80. Audrey Watters, "Education Technology and the New Behaviorism," HackEducation (blog), December 23, 2017, http://hackeducation.com /2017/12/23/top-ed-tech-trends-social-emotional-learning.

81. Michael Walzer, *Spheres of Justice: A Defense of Pluralism and Equality* (New York: Basic Books, 1984).

82. On the power and limits of these utilitarian and deontological theories, see Cohen, *Configuring the Networked Self*. In tension with Western standards of liberalism, Walzer's philosophy has deep resonances with theories of social harmony often associated with traditional East Asian philosophies.

83. Ideally, not just correlations, but also accounts of causation, would matter, too. See, for example, Kiel Brennan-Marquez, "'Plausible Cause': Explanatory Standards in the Age of Powerful Machines," *Vanderbilt Law Review* 70 (2018): 1249–1301. We can all understand why a record of late payments on bills should reduce a credit score; keeping track of them is fair game. Big data factors far beyond one's role as a debtor have less legitimacy in these judgments, however well they may predict repayment problems.

84. Jürgen Habermas, *The Theory of Communicative Action,* vol. 2, *Lifeworld and System: A Critique of Functionalist Reason*, trans. Thomas McCarthy (Boston, MA: Beacon Press, 1987). For an introduction to these ideas, see Hugh Baxter, "System and Lifeworld in Habermas's Theory of Law," *Cardozo Law Review* 23 (2002): 473–615.

85. The irony of Habermas's formulation (dating back to an era when the "expert systems" of AI were clearly interpretable) was that the metaphor of "juridification" turned to the legal field as a particularly opaque and overcomplicated mode of social ordering, while the legal system's basic standard of reason-giving is now a standard of accountability applied to automated systems.

86. This self-instrumentalization is well-critiqued as a species of alienation in Hartmut Rosa, *Resonance: A Sociology of Our Relationship to the World*, trans. James Wagner (Cambridge, UK: Polity Press, 2019). See also Richard M. Re and Alicia Solow-Niederman, "Developing Artificially Intelligent Justice," *Stanford Technology Law Review* 22 (2019): 242, 275, discussing alienation from automated systems.

87.   David Beer, *The Data Gaze: Capitalism, Power and Perception* (Los Angeles: Sage, 2018); Meredith Broussard, *Artificial Unintelligence* (Cambridge, MA: MIT Press, 2018).

88.   Charles Taylor, "Foucault on Freedom and Truth," *Political Theory* 12, no. 2 (1984): 152–183.

89.   William Davies, *Nervous States: Democracy and the Decline of Reason* (New York: Norton, 2018).

90.   Adolph A. Berle, *Power* (New York: Harcourt, 1969); John Gaventa, *Power and Powerlessness: Quiescence and Rebellion in an Appalachian Valley* (Urbana: University of Illinois Press, 1980); Stephen Lukes, *Power* (New York: New York University Press, 1986).

91.   Jonathan Schell, *The Unconquerable World: Power, Nonviolence, and the Will of the People* (New York: Metropolitan Books, 2003).

92.   Elizabeth Warren and Amelia Warren Tyagi, *The Two-Income Trap: Why Middle-Class Mothers and Fathers Are Going Broke* (New York: Basic Books, 2003); Robert Frank, *Falling Behind: How Rising Inequality Harms the Middle Class* (Berkeley: University of California Press, 2007).

93.   Frank Pasquale, "Reclaiming Egalitarianism in the Political Theory of Campaign Finance Reform," *University of Illinois Law Review* (2008): 599–660.

6. AUTONOMOUS FORCES

1.   David Silver, Julian Schrittwieser, Karen Simonyan, Ioannis Antonoglou, Aja Huang, Arthur Guez, Thomas Hubert et al., "Mastering the Game of Go without Human Knowledge," *Nature* 550 (2017): 354–359.

2.   Peter Galison, "The Ontology of the Enemy: Norbert Wiener and the Cybernetic Vision," *Critical Inquiry* 21 (1994): 228–266.

3.   Future of Life Institute, "Slaughterbots," YouTube, November 13, 2017, https://www.youtube.com/watch?v=HipTO_7mUOw; Jessica Cussins, "AI Researchers Create Video to Call for Autonomous Weapons Ban at UN," *Future of Life Institute*, November 14, 2017, https://futureoflife.org/2017/11/14/ai-researchers-create-video-call-autonomous-weapons-ban-un/?cn-reloaded=1.

4.   Shane Harris, "The Brave New Battlefield," *Defining Ideas*, September 19, 2012, http://www.hoover.org/research/brave-new-battlefield; Benjamin

Wittes and Gabriella Blum, *The Future of Violence- Robots and Germs, Hackers and Drones: Confronting the New Age of Threat* (New York: Basic Books, 2015).

5. Gabriella Blum, "Invisible Threats," *Hoover Institution: Emerging Threats,* 2012, https://www.hoover.org/sites/default/files/research/docs/emerging threats_blum.pdf.

6. P. W. Singer and August Cole, *Ghost Fleet: A Novel of the Next World War* (New York: Houghton, 2015).

7. P. W. Singer, "Military Robots and the Future of War," TED talk, February 2009, https://www.ted.com/talks/pw_singer_on_robots_of _war; P. W. Singer, "News and Events," https://www.pwsinger.com/news -and-events/; MCOE Online, "August Cole Discusses Future Fighting Possibilities," YouTube, March 3, 2016, https://www.youtube.com/watch ?v=vl_J9_x-yOk.

8. Anatoly Dneprov, *Crabs on the Island* (Moscow: Mir Publishers, 1968), 10: "'Surely I told you I want to improve my robots.' 'Well, so what? Take your blue prints and work out how to do it. Why this civil war? If this goes on, they will devour each other.' 'Just so. And the most perfect ones will survive.'" Thanks to Michael Froomkin for drawing my attention to the story.

9. Pedro Domingos hypothesizes a Robotic Park, like Jurassic Park, as "a massive robot factory surrounded by ten thousand square miles of jungle, urban and otherwise. . . . I suggested it as a thought experiment at a DARPA [Defense Advanced Research Projects] workshop a few years ago, and one of the military brass present said matter-of-factly, 'That's feasible'"; *The Master Algorithm: How the Quest for the Ultimate Learning Machine Will Remake Our World* (New York: Basic Books, 2015), 121.

10. Bill Joy, "Why the Future Doesn't Need Us," *Wired,* April 1, 2000, https://www.wired.com/2000/04/joy-2/.

11. Brad Turner, "Cooking Protestors Alive: The Excessive-Force Implica- tions of the Active Denial System," *Duke Law & Technology Review* 11 (2012): 332–356.

12. Michael Schmitt, "Regulating Autonomous Weapons Might Be Smarter than Banning Them," *Just Security,* August 10, 2015, https://www .justsecurity.org/25333/regulating-autonomous-weapons-smarter -banning/.

13. Of course, there could be indiscriminate killer robots, set to kill anyone in their path in a particular territory or to kill anyone of a particular appearance. I address the problems these robots cause later in the chapter. They are more akin to the types of advances in destructive capacity that have long been part of the law of war.

14. Ronald Arkin, "The Case for Banning Killer Robots: Counterpoint," *Communications of the ACM* 58, no. 12 (2015): 46–47.

15. Noel E. Sharkey, "The Evitability of Autonomous Robot Warfare," *International Review of the Red Cross* 94 (2012): 787–799. For critical perspectives on substitutive automation of criminal law enforcement personnel, see Elizabeth E. Joh, "Policing Police Robots," *UCLA Law Review Discourse* 64 (2016): 516–543.

16. Thomas Gibbons-Neff, "U.S. Troops Deployed to Bolster Security at American Embassy in South Sudan," *Washington Post*, July 13, 2016, https://www.washingtonpost.com/news/checkpoint/wp/2016/07/13/u-s -troops-deployed-to-bolster-security-at-american-embassy-in-south -sudan/?utm_term=.8825ec86285d; David E. Sanger and Eric Schmitt, "Yemen Withdraws Permission for U.S. Antiterror Ground Missions," *New York Times*, February 7, 2017, https://www.nytimes.com/2017/02/07 /world/middleeast/yemen-special-operations-missions.html.

17. Peter W. Singer, *LikeWar: The Weaponization of Social Media* (Boston: Houghton Mifflin Harcourt, 2018); Peter Pomerantsev, "The Hidden Author of Putinism: How Vladislav Surkov Invented the New Russia," *Atlantic*, November 7, 2014, https://www.theatlantic.com/international /archive/2014/11/hidden-author-putinism-russia-vladislav-surkov/382489/.

18. Alexander Kott, David Alberts, Amy Zalman, Paulo Shakarian, Fernando Maymi, Cliff Wang, and Gang Qu, "Visualizing the Tactical Ground Battlefield in the Year 2050: Workshop Report," *US Army Research Laboratory*, June 2015, https://www.arl.army.mil/arlreports/2015/ARL-SR -0327.pdf; Lucy Ash, "How Russia Outfoxes Its Enemies," *BBC News*, January 29, 2015, http://www.bbc.com/news/magazine-31020283.

19. Rebecca Crootof, "The Killer Robots Are Here," *Cardozo Law Review* 36 (2015): 1837–1915.

20. 1977 Protocol 1 Additional to the Geneva Conventions of August 12, 1949, and Relating to the Protection of Victims of Non-International Armed Conflicts, article 51(3); 1978 Protocol 1 Additional to the Geneva

Conventions of August 12, 1949, and Relating to the Protection of Victims of Non-International Armed Conflicts, article 13(3).

21. Bonnie Docherty, *Losing Humanity: The Case against Killer Robots* (New York: Human Rights Watch, 2012), 31.

22. Mike Crang and Stephen Graham, "Sentient Cities: Ambient Intelligence and the Politics of Urban Space," *Information Communication and Society* 10 (2007): 789, 799 ("The key to this new type of conflict, which profoundly embeds the new 'battlespaces' in urban civilian life, is to mobilize ambient intelligence. Embedded in cities and urban infrastructures, to provide the 'battlespace awareness' necessary to identify, track and target lurking insurgents, terrorists and other 'targets'").

23. Grégoire Chamayou, *A Theory of the Drone* (New York: New Press, 2015), 143–145; Luke A. Whittemore, "Proportionality Decision Making in Targeting: Heuristics, Cognitive Biases, and the Law," *Harvard National Security Journal* 7 (2016): 577, 593 ("At each step, the commander must weigh probabilities to fit something into one category or the other. Couple this uncertainty with the vagueness of the terms, and human decision making under the principle of distinction also is likely to rely on heuristic principles to assess probabilities and predict outcomes").

24. 1977 Protocol 1 Additional to the Geneva Conventions of August 12, 1949, and Relating to the Protection of Victims of Non-International Armed Conflicts, art. 51(5)(b).

25. United States Air Force Judge Advocate General International and Operations Law Division, *Air Force Operations and the Law: A Guide for Air and Space Forces* (Washington, DC: International and Operations Law Division, Judge Advocate General's Department, 2002), 27.

26. U. C. Jha, *Killer Robots: Lethal Autonomous Weapon System Legal, Ethical and Moral Challenges* (New Delhi: Vij Books India Pvt. Ltd., 2016).

27. Peter Asaro, "*Jus Nascendi,* Robotic Weapons, and the Martens Clause," in *Robot Law,* ed. Ryan Calo, Michael Froomkin, and Ian Kerr (Cheltenham, UK: Edward Elgar Publishing, 2016), 377.

28. Ibid., 378. This position finds support in my recent work, "A Rule of Persons, Not Machines: The Limits of Legal Automation," *George Washington Law Review* 79 (2019): 1–55, which describes the shortcomings of automated application of law.

29. Quinta Jurecic, "The Lawfare Podcast: Samuel Moyn on 'How Warfare Became Both More Humane and Harder to End,'" *Lawfare,* October 22,

2016, https://www.lawfareblog.com/lawfare-podcast-samuel-moyn-how
-warfare-became-both-more-humane-and-harder-end.

30. Harold Hongju Koh, "The War Powers and Humanitarian Intervention," *Houston Law Review* 53 (2016): 971–1034.

31. Chamayou, *A Theory of the Drone,* 93, 95.

32. Ibid., 143 ("The fact that your weapon enables you to destroy precisely whomever you wish does not mean that you are more capable of making out who is and who is not a legitimate target").

33. Ibid., 147.

34. The blurring line between offense and defense is also a theme of the cyberwar literature (and has informed cryptography since at least the 1930s). Galison, "Ontology of the Enemy."

35. William Bogard, *The Simulation of Surveillance: Hypercontrol in Telematic Societies* (New York: Cambridge University Press, 1996), 85.

36. Paul Scharre, *Army of None: Autonomous Weapons and the Future of War* (New York: Norton, 2018).

37. Kathleen Belew, *Bring the War Home: The White Power Movement and Paramilitary America* (Cambridge, MA: Harvard University Press, 2018), discussing the popularity of Andrew Macdonald's novel *The Turner Diaries* (Charlottesville, VA: National Vanguard Books, 1999).

38. Brian Massumi, *Ontopower: War, Powers, and the State of Perception* (Durham, NC: Duke University Press, 2006); Joseph A. Schumpeter, *The Economics and Sociology of Capitalism,* (Princeton: Princeton University Press, 1991), 157.

39. P. W. Singer, *Wired for War: The Robotics Revolution and Conflict in the 21st Century* (New York: Penguin Books, 2009), 127.

40. Ibid., 127.

41. Ian Kerr and Katie Szilagyi, "Evitable Conflicts, Inevitable Technologies? The Science and Fiction of Robotic Warfare and IHL," *Law, Culture, and Humanities* 14, no. 1 (2014): 45–82, http://lch.sagepub.com/content /early/2014/01/01/1743872113509443.full.pdf+html (describing how lethal autonomous robots can become a "force multiplier of military necessity.").

42. Ariel Ezrahi and Maurice E. Stucke, *Virtual Competition: The Promise and Perils of the Algorithm-Driven Economy* (Cambridge, MA: Harvard University Press, 2016), 13. Amazon's pricing algorithms made headlines

when they led to an unintended escalation in price of Peter Lawrence's book *The Making of a Fly*. At its peak, Amazon priced the book at $23,698,655.93 (plus $3.99 shipping).

43. Ezrahi and Stucke, *Virtual Competition*, 13.

44. Nathanial Popper, "Knight Capital Says Trading Glitch Cost It $440 Million," *New York Times: DealBook*, August 2, 2012, https://dealbook .nytimes.com/2012/08/02/knight-capital-says-trading-mishap-cost-it-440 -million/.

45. US Securities and Exchange Commission (SEC) and the US Commodity Futures Trading Commission (CFTC), *Findings Regarding the Market Events of May 6, 2010*, September 30, 2010, https://www.sec.gov/news /studies/2010/marketevents-report.pdf.

46. Françoise Hampson, "Military Necessity," in *Crimes of War: What the Public Should Know*, 2nd ed., ed. Roy Guttman, David Rieff, and Anthony Dworkin (New York: Norton, 2007), 297.

47. Protocol Additional to the Geneva Conventions of 12 August 1949, and Relating to the Protection of Victims of International Armed Conflicts (Protocol I), December 12, 1977, 1125 U.N.T.S. 609, Art. 1(2); Rupert Ticehurst, "The Martens Clause and the Laws of Armed Conflict," *International Review of the Red Cross* 79 (1997): 133–142; Theodor Meron, "The Martens Clause, Principles of Humanity, and Dictates of Public Conscience," *American Journal of International Law* 94 (2000): 78–89.

48. Though terms such as "public conscience" may seem to be too vague a concept for even human legal specification, the articulation of foundational public values has already played an important role in constitutional jurisprudence in the United States; see *Roper v. Simmons*, 543 U.S. 551 (2005), which holds that executing individuals who were under eighteen at the time of their capital crimes is barred by the Eighth and Fourteenth Amendments; and *Atkins v. Virginia*, 536 U.S. 304 (2002), which holds that the death penalty is cruel and unusual punishment under the Eighth Amendment in cases involving mentally impaired criminals.

49. Charli Carpenter, "Science Fiction, Popular Culture, and the Concept of Genocide," *Duck of Minerva* (blog), June 10, 2009, http://duckofminerva .com/2009/06/science-fiction-popular-culture-and.html.

50. Campaign to Stop Killer Robots, "About Us," https://www.stopkillerrobots .org/about-us/.

51. For the rest of this chapter, the term "mine" will refer to antipersonnel land mines.

52. Boutros Boutros-Ghali, "The Land Mine Crisis: A Humanitarian Disaster," *Foreign Affairs* 73 (September / October 1994): 8–13.

53. Eric Stover, Allen S. Keller, James Cobey, and Sam Sopheap, "The Medical and Social Consequences of Land Mines in Cambodia," *Journal of the American Medical Association* 272 (1994): 331–336.

54. Ben Wittes and Gabriella Blum, *The Future of Violence: Robots and Germs, Hackers and Drones—Confronting a New Age of Threat* (New York: Basic Books, 2015), 239.

55. Don Huber, "The Landmine Ban: A Case Study in Humanitarian Advocacy," *Thomas J. Watson Jr. Institute for International Studies* 42 (2000), http://www.watsoninstitute.org/pub/op42.pdf.

56. For more on the technologies of tethering and "phoning home," see "Along Came a Spider: The XM-7 RED Mine," *Defense Industry Daily,* August 27, 2013, https://www.defenseindustrydaily.com/along-came-a -spider-the-xm-7-red-04966/; Lester Haines, "Introducing the 'Matrix' Laptop-Triggered Landmine," *Register,* April 12, 2005, https://www.the register.co.uk/2005/04/12/laptop_triggered_landmine/; Brian Bergstein, "'Smart' Land Mines, with Remote Control," *NBC News.com,* April 4, 2004, http://www.nbcnews.com/id/4664710/ns/technology_and _science-science/t/smart-land-mines-remote-control/.

57. Convention on the Prohibition of the Use, Stockpiling, Production and Transfer of Anti-Personnel Mines and on their Destruction, September 18, 1997, 2056 U.N.T.S. 211.

58. Ibid.

59. P. W. Singer, "Military Robots and the Laws of War," *New Atlantis* 23 (2009): 25–45.

60. Ibid., 44–45. Rebecca Crootof has extensively applied tort concepts to the field. Rebecca Crootof, "War Torts: Accountability for Autonomous Weapons," *University of Pennsylvania Law Review* 164 (2016): 1347–1402.

61. See, for example, Samir Chopra and Laurence White, *Legal Theory of Autonomous Artificial Agents* (Ann Arbor: University of Michigan Press, 2011).

62. Dustin A. Lewis, Gabriella Blum, and Naz K. Modirzadeh, "War-Algorithm Accountability," Harvard Law School Program on Interna-

tional Law and Armed Conflict Research Briefing, August 2016, https://papers.ssrn.com/sol3/papers.cfm?abstract_id=2832734. There is already some work on attribution for drones generally. Joseph Lorenzo Hall, "'License Plates' for Drones," *CDT Blog,* March 8, 2013, https://cdt.org/blog/license-plates-for-drones/.

63. See, for example Hall, "License Plates for Drones"; 14 C.F.R. § 48.100 (2016), which requires registration of all small unmanned aircraft other than model aircraft as of August 29, 2016. For the dangers of unattributed drone flights, see A. Michael Froomkin and P. Zak Colangelo, "Self-Defense against Robots and Drones," *Connecticut Law Review* 48 (2015): 1–70.

64. Ellen Nakashima, "Cyber Researchers Confirm Russian Government Hack of Democratic National Committee," *Washington Post,* June 20, 2016, https://www.washingtonpost.com/world/national-security/cyber-researchers-confirm-russian-government-hack-of-democratic-national-committee/2016/06/20/e7375bc0-3719-11e6-9ccd-d6005beac8b3_story.html?utm_term=.90afb7ecbcc6; see also Mandiant Consulting, "M-Trends 2016," February 2016, https://www2.fireeye.com/PPC-m-trends-2016-trends-statistics-mandiant.html.

65. Paul Ohm, "The Fourth Amendment in a World without Privacy," *Mississippi Law Journal* 81 (2012): 1346–1347

66. Paul Scharre, "The False Choice of Humans vs. Automation," *Temple International and Comparative Law Journal* 30 (2016): 151–165.

67. Nick Bilton, "Fake News Is about to Get Even Scarier Than You Ever Dreamed," *Vanity Fair,* January 26, 2017, http://www.vanityfair.com/news/2017/01/fake-news-technology.

68. Paul Ohm and Jonathan Frankel, "Desirable Inefficiency," *Florida Law Review* 70 (2018): 777–838. The concept of a "garrison state" was developed in Harold D. Lasswell, "The Garrison State," *American Journal of Sociology* 46, no. 4 (1941): 455–468.

69. Stockholm International Peace Research Institute, "World Military Expenditure Grows to $1.8 Trillion in 2018," *Sipri,* April 29, 2019, https://www.sipri.org/media/press-release/2019/world-military-expenditure-grows-18-trillion-2018.

70. Christopher A. Preble, *The Power Problem: How American Military Dominance Makes Us Less Safe, Less Prosperous, and Less Free* (Ithaca,

NY: Cornell University Press, 2009); John Mueller and Mark Stewart, *Chasing Ghosts: The Policing of Terrorism* (Oxford: Oxford University Press, 2015).

71. Moisi is quoted in Katrin Bennhold, "'Sadness' and Disbelief from a World Missing American Leadership, *New York Times*, April 23, 2020.

72. Andrew Bacevich, *The Age of Illusions: How America Squandered Its Cold War Victory* (New York: Metropolitan Books, 2020); Nikhil Pal Singh, Quincy Institute for Responsible Statecraft, "Enough Toxic Militarism," *Quincy Brief*, December 4, 2019, https://quincyinst.org/2019/12/04/quincy-brief-enough-toxic-militarism/.

73. T.X. Hammes, "The Future of Warfare: Small, Many, Smart vs. Few & Exquisite?," *War on the Rocks*, July 16, 2014, https://warontherocks.com/2014/07/the-future-of-warfare-small-many-smart-vs-few-exquisite/.

74. Sulmaan Wasif Khan, *Haunted by Chaos: China's Grand Strategy from Mao Zedong to Xi Jinping* (Cambridge, MA: Harvard University Press, 2018), 232. Xi's "use of military power has undermined the larger strategic goal it was meant to achieve: a stable neighborhood. Countries like Japan and Vietnam have not backed down when threatened by fishing militias and aircraft carriers. Instead, China's forcefulness seems to have spurred their quests for military power."

75. Henry Farrell, "Seeing Like a Finite State Machine," *Crooked Timber* (blog), November 25, 2019, http://crookedtimber.org/2019/11/25/seeing-like-a-finite-state-machine/.

76. Robert A. Burton, "Donald Trump, Our A.I. President," *New York Times*, May 22, 2017, https://www.nytimes.com/2017/05/22/opinion/donald-trump-our-ai-president.html.

77. John Glaser, Christopher H. Preble, and A. Trevor Thrall, *Fuel to the Fire: How Trump Made America's Broken Foreign Policy Even Worse (and How We Can Recover)* (Washington, DC: Cato Institute, 2019).

78. Bernard Harcourt, *Critique and Praxis* (New York: Columbia University Press, 2020).

79. Etel Solingen, *Nuclear Logics: Contrasting Paths in East Asia and the Middle East* (Princeton: Princeton University Press, 2007), 253.

80. Noam Schreiber and Kate Conger, "The Great Google Revolt," *New York Times Magazine*, February 18, 2020, https://www.nytimes.com/interactive/2020/02/18/magazine/google-revolt.html; Letter from Google Employees

to Sundar Pichai, Google Chief Executive, 2018, https://static01.nyt.com /files/2018/technology/googleletter.pdf; Scott Shane and Daisuke Wakabayashi, "'The Business of War': Google Employees Protest Work for the Pentagon," *New York Times,* April 4, 2018, https://www.nytimes.com/2018 /04/04/technology/google-letter-ceo-pentagon-project.html.

81. Kate Conger, "Google Removes 'Don't Be Evil' Clause from Its Code of Conduct," *Gizmodo,* May 18, 2018, https://gizmodo.com/google-removes -nearly-all-mentions-of-dont-be-evil-from-1826153393.

82. Tom Upchurch, "How China Could Beat the West in the Deadly Race for AI Weapons," *Wired,* August 8, 2018, https://www.wired.co.uk/article /artificial-intelligence-weapons-warfare-project-maven-google-china.

83. Elsa B Kania, "Chinese Military Innovation in the AI Revolution," *RUSI Journal* 164, nos. 5–6 (2019): 26–34. "In Tianjin, a new AI Military-Civil Fusion Innovation Center . . . , located next to the National Supercomputer Center, was established by the local government in partnership with the Academy of Military Science."

84. Ashwin Acharya and Zachary Arnold, *Chinese Public AI R&D Spending: Provisional Findings* (Washington, DC: Center for Security and Emerging Technology, 2019), https://cset.georgetown.edu/wp-content/uploads /Chinese-Public-AI-RD-Spending-Provisional-Findings-1.pdf. Acharya and Arnold found China's military AI spending to be lower than previously estimated.

85. Luciano Floridi and Mariarosaria Taddeo, eds., *The Ethics of Information Warfare* (Cham, CH: Springer, 2014); Singer and Brooking, *LikeWar.*

86. Samuel Bowles and Arjun Jayadev, "Guard Labor: An Essay in Honor of Pranab Bardhan," University of Massachusetts Amherst Working Paper 2004–2015 (2004), https://scholarworks.umass.edu/econ_workingpaper /63/; Harold Lasswell, "The Garrison State," *American Journal of Sociology* 46 (1941): 455–468.

87. Ryan Gallagher, "Cameras Linked to Chinese Government Stir Alarm in U.K. Parliament," *Intercept,* April 9, 2019, https://theintercept.com/2019 /04/09/hikvision-cameras-uk-parliament/; Arthur Gwagwa, "Exporting Repression? China's Artificial Intelligence Push into Africa," *Council on Foreign Relations: Net Politics* (blog), December 17, 2018, https://www .cfr.org/blog/exporting-repression-chinas-artificial-intelligence-push -africa.

88. "Companies Involved in Expanding China's Public Security Apparatus in Xinjiang," *ChinAI Newsletter #11,* trans. Jeff Ding, May 21, 2018, https://

chinai.substack.com/p/chinai-newsletter-11-companies-involved-in
-expanding-chinas-public-security-apparatus-in-xinjiang.

89. Doug Bandow, "The Case for a Much Smaller Military," *Fortune* 135, no. 12 (1997): 25–26.

90. Bernard Harcourt, *The Counterrevolution: How Our Government Went to War against Its Own Citizens* (New York: Public Books, 2018).

91. Sina Najafi and Peter Galison, "The Ontology of the Enemy: An Interview with Peter Galison," *Cabinet Magazine,* Fall / Winter 2003, http:// cabinetmagazine.org/issues/12/najafi2.php.

92. William E. Connolly, *The Fragility of Things: Self-Organizing Processes, Neoliberal Fantasies, and Democratic Activism* (Durham, NC: Duke University Press, 2013).

93. Rosa Brooks, *How Everything Became War and the Military Became Everything: Tales from the Pentagon* (New York: Simon & Schuster, 2017).

94. Deborah Brautigam, *The Dragon's Gift: The Real Story of China in Africa* (New York: Oxford University Press, 2009). On the category of the "investee," see Michael Feher, *Rated Agency: Investee Politics in a Speculative Age*, trans. Gregory Elliott (New York: Zone Books, 2018).

95. For example, while public health experts warned the George W. Bush Administration that "the best way to manage bioterrorism is to improve the management of existing public health threats," officials instead "fast-tracked vaccination programs for smallpox and anthrax, based on fanciful scenarios that might have embarrassed Tom Clancy." Mike Davis, *The Monster at Our Door* (New York: The New Press, 2005), 135. See also Amy Kapczynski and Gregg Gonsalves, "Alone against the Virus," *Boston Review*, March 13, 2020, http://bostonreview.net/class-inequality-science -nature/amy-kapczynski-gregg-gonsalves-alone-against-virus.

96. Ian G. R. Shaw, *Predator Empire: Drone Warfare and Full Spectrum Dominance* (Minneapolis: University of Minnesota Press, 2016).

7. RETHINKING THE POLITICAL ECONOMY OF AUTOMATION

1. Law and political economy scholarship advances this methodological innovation. Jedediah S. Britton-Purdy, David Singh Grewal, Amy Kapczynski, and K. Sabeel Rahman, "Building a Law-and-Political- Economy Framework: Beyond the Twentieth-Century Synthesis," *Yale Law Journal* 130 (2020): 1784–1835.

2. Martha McCluskey, "Defining the Economic Pie, Not Dividing or Maximizing It," *Critical Analysis of Law* 5 (2018): 77–98.

3. The term "local" here is polysemic, denoting the sense of both geographical nearness to the AI / robotics implementation and functional domain expertise. Clifford Geertz, *Local Knowledge: Further Essays in Interpretive Anthropology* (New York: Basic Books, 1983).

4. For more on how this can be done in the manufacturing and service sectors not focused on in this book, see Roberto Unger, *The Knowledge Economy* (New York: Verso, 2019).

5. Roger Boesche, "Why Could Tocqueville Predict So Well?" *Political Theory* 11 (1983): 79–103.

6. Joseph E. Aoun, *Robot-Proof: Higher Education in the Age of Artificial Intelligence* (Cambridge, MA: MIT Press, 2017).

7. I borrow the term "most advanced modes of production" from Roberto Unger, *The Knowledge Economy*.

8. Eliza Mackintosh, "Finland Is Winning the War on Fake News. What It's Learned May Be Crucial to Western Democracy," CNN, May 2019, https://www.cnn.com/interactive/2019/05/europe/finland-fake-news-intl/.

9. For example, there is an excellent discussion of the Facebook reaction buttons in David Auerbach, *Bitwise: My Life in Code* (New York: Random House, 2018).

10. Luke Stark, "Algorithmic Psychometrics and the Scalable Subject," *Social Studies of Science* 48, no. 2 (2018): 204–231.

11. Avner Offer, *The Challenge of Affluence: Self-Control and Well-Being in the United States and Britain since 1950* (Oxford: Oxford University Press, 2006); Avner Offer, Rachel Pechey, and Stanley Ulijaszek, eds., *Inequality, Insecurity, and Obesity in Affluent Societies* (Oxford: Oxford University Press, 2012).

12. Michael Simkovic, "The Knowledge Tax," *University of Chicago Law Review* 82 (2015): 1981–2043. Standard budget models routinely fail to reflect these contributions. Michael Simkovic, "Biased Budget Scoring and Underinvestment," *Tax Notes Federal* 166, no. 5 (2020): 75–765.

13. The IEEE Global Initiative on Ethics of Autonomous and Intelligent Systems, *Ethically Aligned Design: A Vision for Prioritizing Human Well-Being with Autonomous and Intelligent Systems*, 1st. ed. (2019), 102.

14. Robert Lee Hale, *Freedom through Law* (New York: Columbia University Press, 1952).

15. Brishen Rogers, "The Law and Political Economy of Workplace Technological Change," *Harvard Civil Rights-Civil Liberties Law Review* 55 (forthcoming 2020).

16. Elizabeth Anderson, *Private Government: How Employers Rule Our Lives (And Why We Don't Talk about It)* (Princeton: Princeton University Press, 2018).

17. Veena B. Dubal, "Wage Slave or Entrepreneur?: Contesting the Dualism of Legal Worker Identities," *California Law Review* 105 (2017): 101–159; Sanjukta Paul, "Uber as For-Profit Hiring Hall: A Price-Fixing Paradox and Its Implications," *Berkeley Journal of Employment and Labor Law* 38 (2017): 233–264.

18. For more responses to critiques of professional organizations, see Sandeep Vaheesan and Frank Pasquale, "The Politics of Professionalism," *Annual Review of Law & Social Science* 14 (2018): 309–327.

19. The humanities and social sciences are critical to exploring such values. Religion also plays a role here, and religious institutions occasionally comment directly on developments in AI and robotics. For an example, see Pontifical Academy of Sciences and Pontifical Academy of Social Sciences, *Final Statement from the Conference on Robotics, AI and Humanity, Science, Ethics and Policy* (2019), http://www.academyof sciences.va/content/accademia/en/events/2019/robotics/statement robotics.html.

20. Charles Duhigg, "Did Uber Steal Google's Intellectual Property?," *New Yorker*, October 15, 2018, https://www.newyorker.com/magazine/2018/10 /22/did-uber-steal-googles-intellectual-property.

21. Ben Green, *The Smart Enough City: Putting Technology in Its Place to Reclaim Our Urban Future* (Cambridge, MA: MIT Press, 2019).

22. James Kwak, *Economism: Bad Economics and the Rise of Inequality* (New York: Pantheon Books, 2017).

23. John Lanchester "Good New Idea" *London Review of Books* 41, no. 14 (2019), https://www.lrb.co.uk/the-paper/v41/n14/john-lanchester/good -new-idea.

24. Hartmut Rosa, *Resonance: A Sociology of Our Relationship to the World*, trans. James Wagner (Medford, MA: Polity Press, 2019).

25. Cristóbal Orrego, "The Universal Destination of the World's Resources," in *Catholic Social Teaching*, eds. Gerard V. Bradley and E. Christian Brugger (Cambridge, UK: Cambridge University Press, 2019), 267–299.

26. Gar Alperovitz and Lew Daly, *Unjust Desserts: How the Rich Are Taking Our Common Inheritance* (New York: New Press, 2010); Guy Standing, *Basic Income: A Guide for the Open Minded* (New Haven: Yale University Press, 2017).

27. Daren Acemoglu and Pascual Restrepo, "Automation and New Tasks: How Technology Displaces and Reinstates Labor," *Journal of Economic Perspectives* 33, no. 2 (2019): 3, 25.

28. Elizabeth Warren United States Senator for Massachusetts, "Senator Warren Unveils Proposal to Tax Wealth of Ultra-Rich Americans," press release, January 24, 2019, https://www.warren.senate.gov/newsroom/press-releases/senator-warren-unveils-proposal-to-tax-wealth-of-ultra-rich-americans.

29. "Tax on Extreme Wealth," Issues, Friends of Bernie Sanders, accessed May 13, 2020, https://berniesanders.com/issues/tax-extreme-wealth/.

30. Karl Widerquist, "The Cost of Basic Income: Back-of-the-Envelope Calculations," *Basic Income Studies* 12, no. 2 (2017): 1–13.

31. Thomas Piketty, *Capital and Ideology* (Cambridge, MA: Harvard University Press, 2020).

32. Paul Fain, "Huge Budget Cut for the University of Alaska," *Inside Higher Ed*, June 29, 2019, https://www.insidehighered.com/quicktakes/2019/06/29/huge-budget-cut-university-alaska.

33. For a vital resource, see Ganesh Sitaraman and Anne L. Alstott, *The Public Option: How to Expand Freedom, Increase Opportunity, and Promote Equality* (Cambridge, MA: Harvard University Press, 2019).

34. Pavlina Tcherneva in conversation with David Roberts, "30 Million Americans are Unemployed. Here's How to Employ Them," *Vox*, May 4, 2020, at https://www.vox.com/science-and-health/2020/5/4/21243725/coronavirus-unemployment-cares-act-federal-job-guarantee-green-new-deal-pavlina-tcherneva. Pavlina Tcherneva, "The Job Guarantee: Design, Jobs, and Implementation," Levy Economics Institute, Working Papers Series No. 902 (2018), https://papers.ssrn.com/sol3/papers.cfm?abstract_id=3155289.

35. William J. Baumol and William G. Bowen, *Performing Arts: The Economic Dilemma* (New York: Twentieth Century Fund, 1966).

36. Carolyn Demitri, Anne Effland, and Neilson Conklin, "The 20th Century Transformation of U.S. Agriculture and Farm Policy," *USDA Economic Information Bulletin* 3 (June 2005): 1, 6, https://www.ers.usda.gov

/webdocs/publications/eib3/13566_eib3_1_.pdf?v=41055; John Seabrook, "The Age of Robot Farmers," *New Yorker,* April 8, 2019, https://www .newyorker.com/magazine/2019/04/15/the-age-of-robot-farmers.

37. Ian D. Wyatt and Daniel E. Hecker, "Occupational Changes during the 20th Century," *Monthly Labor Review* (2006): 35, 55. For example, farming constituted 33 percent of the U.S. labor force in 1910 and only 1.2 percent by 2000. Donald M. Fisk, "American Labor in the 20th Century," *Monthly Labor Review: US Bureau of Labor Statistics,* September 2001, https:// www.bls.gov/opub/mlr/2001/article/american-labor-in-the-20th-century .htm. Goods-producing industries (mining, manufacturing, and construction) made up 31 percent of the U.S. workforce in 1900 and 19 percent in 1999.

38. Jack Balkin, "Information Fiduciaries and the First Amendment," *UC Davis Law Review* 49 (2016): 1183–1234. Balkin primarily aims at imposing certain fiduciary duties on the firms and persons that develop and sell AI systems, including robotics.

39. Natasha Dow Schüll, *Addiction by Design: Machine Gambling in Law Vegas* (Princeton: Princeton University Press, 2012).

40. Goss v. Lopez, 419 U.S. 565 (1975).

41. Frank Pasquale, "Synergy and Tradition: The Unity of Research, Service, and Teaching in Legal Education," *Journal of the Legal Profession* 40 (2015): 25–48.

42. David Stark, *The Sense of Dissonance: Accounts of Worth in Economic Life* (Princeton: Princeton University Press, 2011).

43. Harold J. Wilensky, "The Professionalization of Everyone?" *American Journal of Sociology* 70, no. 2 (1964): 137–158.

44. Mark Blyth, "End Austerity Now," *Project Syndicate,* August 20, 2013, https://www.project-syndicate.org/commentary/why-austerity-is-the -wrong-prescription-for-the-euro-crisis-by-mark-blyth?barrier =accessreg.

45. Phil Wahba and Lisa Baertlein, "McDonald's, Walmart Feel the Pinch As Low-Income Shoppers Struggle in the Slow Recovery," *Huffington Post,* August 16, 2013, http://www.huffingtonpost.com/2013/08/16/mcdonalds -walmart-low-income-shoppers_n_3765489.html.

46. Matthew Yglesias, "The Skills Gap Was a Lie," *Vox,* January 9, 2019, https://www.vox.com/2019/1/7/18166951/skills-gap-modestino-shoag -ballance.

47. National Science and Technology Council, Committee on Technology, *Preparing for the Future of Artificial Intelligence* (Executive Office of the President, October 2016), https://www.whitehouse.gov/sites/default/files /whitehouse_files/microsites/ostp/NSTC/preparing_for_the_future_of _ai.pdf?mc_cid=b2a6abaa55&mc_eid=aeb68cfb98; National Science and Technology Council, Networking and Information Technology Research and Development Subcommittee, *The National Artificial Intelligence Research and Development Strategic Plan* (Executive Office of the President, October 2016), https://www.whitehouse.gov/sites /default/files/whitehouse_files/microsites/ostp/NSTC/national_ai _rd_strategic_plan.pdf.

48. AI Now, *The AI Now Report: The Social and Economic Implications of Artificial Intelligence Technologies in the Near-Term*, September 22, 2016, https://ainowinstitute.org/AI_Now_2016_Report.pdf.

49. Dan Diamond, "Obamacare: The Secret Jobs Program," *Politico,* July 13, 2016), http://www.politico.com/agenda/story/2016/07/what-is-the-effect -of-obamacare-economy-000164.

50. Alyssa Battistoni, "Living, Not Just Surviving," *Jacobin Magazine,* August 2017, https://jacobinmag.com/2017/08/living-not-just-surviving/.

51. Eamonn Fingleton, *In Praise of Hard Industries: Why Manufacturing, Not the Information Economy, Is the Key to Future Prosperity* (New York: Houghton Mifflin, 1999); Andy Grove, "How America Can Create Jobs," *Bloomberg* (July 1, 2010), https://www.bloomberg.com/news/articles/2010 -07-01/andy-grove-how-america-can-create-jobs.

52. Douglas Wolf and Nancy Folbre, eds., *Universal Coverage of Long-Term Care in the United States Can We Get There from Here?* (New York: Russell Sage Foundation, 2012).

53. Nancy Folbre, *For Love and Money: Care Provision in the United States* (New York: Russell Sage Foundation, 2012).

54. Jason Furman, "Is This Time Different? The Opportunities and Challenges of AI," Experts Workshop AI Now Institute 2016 Symposium, July 7, 2016, video, 32:39, https://ainowinstitute.org/symposia/videos/is -this-time-different-the-opportunities-and-challenges-of-ai.html.

55. Such valuation is expertly accomplished in Amalavoyal V. Chari, John Engberg, Kristin N. Ray, and Ateev Mehrotra, "The Opportunity Costs of Informal Elder-Care in the United States: New Estimates from the American Time Use Survey," *Health Services Research* 50 (2015): 871–882.

56. Ibid.

57. Baumol, *The Cost Disease*, 182.

58. LaRue Allen and Bridget B. Kelly, eds, *Transforming the Workplace for Children Birth through Age 8: A Unifying Foundation* (Washington, DC: Institute of Medicine and National Research Council of the National Academies, 2015).

59. David S. Fallis and Amy Brittain, "In Virginia, Thousands of Day-Care Providers Receive No Oversight," *Washington Post,* August 30, 2014, https://www.washingtonpost.com/sf/investigative/2014/08/30/in-virginia -thousands-of-day-care-providers-receive-no-oversight/?hpid=z3&tid=a _inl&utm_term=.0a5aff61e742.

60. Stephanie Kelton, *The Deficit Myth: Modern Monetary Theory and the Birth of the People's Economy* (New York: Public Affairs, 2020).

61. L. Randall Wray, *Modern Monetary Theory*, 3d ed. (New York: Palgrave MacMillan, 2012).

62. Mariana Mazzucato, *The Entrepreneurial State: Debunking Public vs. Private Sector Myths* (London: Penguin, 2018).

63. Kate Aronoff, Alyssa Battistoni, Daniel Aldana Cohen, and Thea Riofrancos, *A Planet to Win: Why We Need a Green New Deal* (New York: Verso, 2019).

64. Gregg Gonsalves and Amy Kapczynski, "The New Politics of Care," *Boston Review* (2020), http://bostonreview.net/politics/gregg-gonsalves -amy-kapczynski-new-politics-care.

65. For insightful comments on this facet of an emergent MMT research agenda, see Nathan Tankus, "Are General Price Level Indices Theoretically Coherent?," http://www.nathantankus.com/notes/are-general-price -level-indices-theoretically-coherent.

66. Kurt Vonnegut, *Player Piano* (New York: Dial Press, 2006).

67. E. M. Forster, "The Machine Stops," *Oxford and Cambridge Review* (1909).

68. Perhaps someday we will speak of the "no cost disease," expertly dissected as a cause of so much of what ails contemporary media. Chris Jay Hoofnagle and Jan Whittington, "Free: Accounting for the Costs of the Internet's Most Popular Price," *University of California at Los Angeles Law Review* 61 (2014): 606–670.

69. Jeff Spross, "How Robots Became a Scapegoat for the Destruction of the Working Class," *Week,* April 29, 2019, https://theweek.com/articles /837759/how-robots-became-scapegoat-destruction-working-class.

70.   Daniel Akst, "What Can We Learn from Past Anxiety over Automation?" *Wilson Quarterly,* Summer 2013, http://wilsonquarterly.com/quarterly /summer-2014-where-have-all-the-jobs-gone/theres-much-learn-from -past-anxiety-over-automation/.

71.   David H. Autor, "Polanyi's Paradox and the Shape of Employment Growth," NBER Working Paper No. 20485 (September 2014), http://www .nber.org/papers/w20485.

8. COMPUTATIONAL POWER AND HUMAN WISDOM

1.   For example, see Stephen Cave, Kanta Dihal, and Sarah Dillon, eds., *AI Narratives: A History of Imaginative Thinking about Intelligent Machines* (Oxford: Oxford University Press, 2020).

2.   Peter Frase writes that "science fiction is to futurism what social theory is to conspiracy theory: an altogether richer, more honest, and more humble enterprise." He writes in a new genre—"social science fiction"—meant to ground the speculation of sci-fi in the empirics of sociology. Peter Frase, *Four Futures: Life after Capitalism* (New York: Verso, 2016), 27. See also William Davies, ed., *Economic Science Fictions* (London: Goldsmiths Press, 2018); Manu Saadia, *Trekonomics: The Economics of Star Trek* (Oakland, CA: Pipertext, 2016); William Bogard, *The Simulation of Surveillance* (Cambridge, UK: Cambridge University Press, 1992).

3.   Richard H. Brown helpfully catalogs such overarching terms of moral and ontological orientation. Richard H. Brown, *A Poetic for Sociology: Toward a Logic of Discovery for the Human Sciences* (New York: Cambridge University Press, 1977), 125–126.

4.   Chantal Mouffe and Nancy Fraser offer compelling descriptions (and critiques) of these rhetorical positionings. Chantal Mouffe, *The Return of the Political* (New York: Verso, 2019); Nancy Fraser, *The Old Is Dying and the New Cannot Be Born* (New York: Verso, 2020).

5.   Frank Pasquale, "To Replace or Respect: Futurology as if People Mattered," *Boundary2 Online,* January 20, 2015, https://www.boundary2.org /2015/01/to-replace-or-respect-futurology-as-if-people-mattered/.

6.   On deep stories, see Arlie Hochschild, *Strangers in Their Own Land: Anger and Mourning on the American Right* (New York: New Press, 2016). On basic narrative forms, see George Lakoff, *Thinking Points: Communi-*

cating Our American Values and Vision (New York: Farrar, Straus and Giroux, 2006).

7. Clifford Geertz, *The Interpretation of Cultures* (New York: Basic Books, 1973), 89.

8. For a compelling analysis of robotics in popular culture, see Robert M. Geraci, *Apocalyptic AI: Visions of Heaven in Robotics, Artificial Intelligence, and Virtual Reality* (New York: Oxford University Press, 2010). Wikipedia features a dynamically updated account of "AI in Film": Wikipedia, "List of Artificial Intelligence in Films," https://en.wikipedia .org/wiki/List_of_artificial_intelligence_films. Unsurprisingly, there is no rival "List of Intelligence Augmentation in Films."

9. Langdon Winner, *Autonomous Technology: Technics-out-of-Control as a Theme in Political Thought* (Cambridge, MA: MIT Press, 1977).

10. Such approaches are probably already in use by advertisers, who have been reported to utilize images that look like their viewer in order to cultivate a positive response.

11. Yuval Noah Harari, "The Rise of the Useless Class," *Ideas.Ted.com*, February 24, 2017, https://ideas.ted.com/the-rise-of-the-useless-class/.

12. Douglas McCollam, "It Is What It Is . . . But What Is It?" *Slate,* February 15, 2008, https://slate.com/culture/2008/02/it-is-what-it-is-a-sports -cliche-for-our-times.html.

13. A good recent overview is Nicholas Weaver, "Inside Risks of Cryptocurrencies," *Communications of the ACM* 61, no. 6 (2018): 1–5. For broader background, see David Golumbia, *The Politics of Bitcoin: Software as Right-Wing Extremism* (Minneapolis: University of Minnesota Press, 2016).

14. This possibility is expertly critiqued by Angela Walch; see "The Path of the Blockchain Lexicon (and the Law)," *Review of Banking and Financial Law* 36 (2017): 713–765.

15. Evan Osnos, "Doomsday Prep for the Super-Rich," *New Yorker,* January 22, 2017, https://www.newyorker.com/magazine/2017/01/30 /doomsday-prep-for-the-super-rich.

16. Noah Gallagher Shannon, "Climate Chaos Is Coming—and the Pinkertons Are Ready," *New York Times Magazine,* April 10, 2019, https://www .nytimes.com/interactive/2019/04/10/magazine/climate-change -pinkertons.html.

17.   Douglas Rushkoff, "Survival of the Richest," *Medium,* July 5, 2018, https://medium.com/s/futurehuman/survival-of-the-richest -9ef6cddd0cc1; see also Bill McKibben, *Falter: Has the Human Game Begun to Play Itself Out?* (New York: Henry Holt, 2019).

18.   Frank Pasquale, "Paradoxes of Privacy in an Era of Asymmetrical Social Control," in Aleš Završnik, *Big Data, Crime, and Social Control* (New York: Routledge, 2018), 31–57.

19.   Richard Susskind and Daniel Susskind, *The Future of the Professions: How Technology Will Transform the Work of Human Experts* (New York: Oxford University Press, 2016): 247.

20.   W. Olaf Stapledon, *Last and First Men: A Story of the Near and Far Future* (Mineola, NY: Dover Publications, 1931).

21.   Devin Fidler, "Here's How Managers Can Be Replaced by Software," *Harvard Business Review,* April 21, 2015, https://hbr.org/2015/04/heres -how-managers-can-be-replaced-by-software.

22.   RAND Corp., "Skin in the Game: How Consumer-Directed Plans Affect the Cost and Use of Health Care," Research Brief no. RB-9672, 2012, http://www.rand.org/pubs/research_briefs/RB9672/index1.html.

23.   Alison Griswold, "Uber Drivers Fired in New York Can Now Appeal before a Panel of Their Peers," *Quartz,* November 23, 2016, https://qz.com /843967/uber-drivers-fired-in-new-york-can-now-appeal-before-a-panel -of-their-peers/.

24.   Trebor Scholz and Nathan Schneider, *Ours to Hack and to Own: The Rise of Platform Cooperativism* (New York: OR Books, 2017).

25.   Ian McEwan, *Machines Life Me: A Novel* (New York: Doubleday, 2019), 4.

26.   Ibid., 65.

27.   Erik Gray, "Machines Like Me, But They Love You," *Public Books,* August 12, 2019, https://www.publicbooks.org/machines-like-me-but -they-love-you/.

28.   McEwan, *Machines Like Me*, 161–162.

29.   As Mark Andrejevic astutely observes, "Automation embraces the logic of immediation (the invisibility or disappearance of the medium) that parallels the promise of virtual reality. . . . This is the promise of machine 'language'—which differs from human language precisely because it is non-representational. For the machine, there is no space between sign and referent: there is no 'lack' in a language that is complete unto itself.

In this respect, machine language is 'psychotic' . . . [envisioning] the perfection of social life through its obliteration." Andrejevic, *Automated Media*, 72.

30. Richard Powers, *Galatea 2.2* (New York: Picador, 1995).

31. Cixin Liu, *The Three-Body Problem*, trans. Ken Liu (New York: Tor, 2006). A similarly immature account of transparency as a route to human perfectibility was articulated in a long tribute to Elon Musk's Neuralink. Tim Urban, "Neuralink and the Brain's Magical Future," *Wait but Why*, April 20, 2017, https://waitbutwhy.com/2017/04/neuralink.html.

32. Abeba Birhane and Jelle van Dijk, "Robot Rights? Let's Talk about Human Welfare Instead," *arXiv*, January 14, 2020, https://arxiv.org/abs /2001.05046.

33. McEwan, *Machines Like Me*, 195.

34. As of April 20, 2020, "impeccably eleemosynary" has never appeared in any other work indexed on Google. More readers would instantly grasp the main thrust of the meaning if I had used the words "perfectly charitable." But the words I chose have certain resonances lost in the simpler phrasing. If I were reading this book aloud, I would in part convey those resonances by saying "impeccably eleemosynary" with an undertone of sarcasm and savor, as if to underscore a certain pleasure in (and distance from) this assessment of Adam's actions. The behavior is commendable, but what is the spirit behind it—if any? That moral uncertainty is rooted in old debates over the relative value of faith and works, intent and results. I try, if clumsily, to reflect those normative tensions with the word "impeccable," which is meant to bring to mind the Latin *peccavi*, a confession of sin with some hope of absolution. Eleemosynary is an unusual (and even antiquated) word meant to connote the strangeness (and even primitiveness) of Adam's actions. For more on the need to preserve these irregularities and ambiguities of language, see my discussion of Lawrence Joseph's poem "Who Talks Like That?," in Frank Pasquale, "The Substance of Poetic Procedure," *Law & Literature* (2020): 1–46. I see no way, even in principle, for computational systems to fully encompass the shades of meaning expressed here, let alone that which is adumbrated and unexpressed.

35. McEwan, *Machines Like Me*, 299.

36. Ibid., 299.

37. Ibid., 301.

38. Ibid., 303.

39. Ibid., 304.

40. Ibid., 329.

41. Ibid., 330.

42. Ludwig Wittgenstein, *Philosophical Investigations*, trans. G. E. M. Anscombe, P. M. S. Hacker, and Joachim Schulte, 4th ed. (West Sussex, UK: Blackwell, 2009), sec. 43.

43. For a fictional account of potentially time-limited digital entities with personalities and intelligence, see Ted Chiang, *The Lifecycle of Software Objects* (Burton, MI: Subterranean Press, 2010).

44. Hubert L. Dreyfus, *What Computers Still Can't Do*, rev. ed. (Cambridge, MA: MIT Press, 1992). "My thesis, which owes a lot to Wittgenstein, is that whenever human behavior is analyzed in terms of rules, these rules must always contain a ceteris paribus condition, i.e., they apply 'everything else being equal,' and what 'everything else' and 'equal' means in any specific situation can never be fully spelled out without a regress. Moreover, this ceteris paribus condition is not merely an annoyance. . . . Rather the ceteris paribus condition points to a background of practices which are the condition of the possibility of all rulelike activity. In explaining our actions we must always sooner or later fall back on our everyday practices and simply say 'this is what we do' or 'that's what it is to be a human being.' Thus in the last analysis all intelligibility and all intelligent behavior must be traced back to our sense of what we are, which is, according to this argument, necessarily, on pain of regress, something we can never explicitly know." Ibid., 56–57. In other words, "since intelligence must be situated it cannot be separated from the rest of human life." Ibid., 62.

45. Meredith Broussard, *Artificial Unintelligence: How Computers Misunderstand the World* (Cambridge, MA: MIT Press, 2018), 7–9, 75.

46. Carrie Goldberg, *Nobody's Victim: Fighting Psychos, Stalkers, Pervs, and Trolls* (New York: Plume, 2019).

47. Sasha Costanza-Chock, *Design Justice: Community-Led Practices to Build the Worlds We Need* (Cambridge, MA: MIT Press, 2020).

48. George Lakoff and Mark Johnson, *Philosophy in the Flesh: The Embodied Mind and the Challenge to Western Thought* (New York: Basic Books, 1999), 4.

49. Nick Bostrom, *Superintelligence: Paths, Dangers, Strategies* (Oxford: Oxford University Press, 2014), 134–135.

50. For examples of such shortcomings, see Rana Foroohar, *Don't Be Evil: How Big Tech Betrayed Its Founding Principles—And All of Us* (New York: Currency, 2019); Amy Webb, *The Big Nine: How the Tech Titans and Their Thinking Machines Could Warp Humanity* (New York: Public Affairs, 2019).

51. Byron Reese categorizes these metaphysical questions with great acuity in *The Fourth Age: Smart Robots, Conscious Computers, and the Future of Humanity* (New York: Simon and Schuster, 2018), 15–54.

52. Paul Dumouchel and Luisa Damiano, *Living with Robots*, trans. Malcolm DeBevoise (Cambridge, MA: Harvard University Press, 2017), 21–23, 167–169.

53. Serge Tisseron's objections to mechanized emotion as a "programmed illusion" also apply here. Catherine Vincent, "Serge Tisseron: 'Les Robots vont Modifier la Psychologie Humaine,'" *Le Monde*, July 12, 2018, https://www.lemonde.fr/idees/article/2018/07/12/serge-tisseron-les-robots-vont-modifier-la-psychologie-humaine_5330469_3232.html.

54. True, a human could be immediately anesthetized, and perhaps medicine will advance to ensure the safe and ready application of such pain killing at some point in the future. But even if such a desirable short-term palliative is possible, it is hard to imagine it continually applied throughout one's recovery from injury, given the ways in which pain can function as an indicator of failures to heal.

55. Kiel Brennan-Marquez and Stephen E. Henderson, "Artificial Intelligence and Role-Reversible Judgment," *The Journal of Criminal Law and Criminology* 109, no. 2 (2019): 137–164.

56. Matthew Lopez, *The Inheritance* (London: Faber, 2018), 8.

57. For a literary perspective on these two possibilities, see Walker Percy, "The Man on the Train," in *Message in a Bottle: How Queer Man Is, How Queer Language Is, and What One Has to Do with the Other* (New York: Picador USA, 2000), 83–100.

58. Gert J. J. Bietsa, *The Beautiful Risk of Education* (London: Routledge, 2016).

59. Brett Frischmann and Evan Selinger, *Re-engineering Humanity* (Cambridge: Cambridge University Press, 2018), 269–295.

60. Frank Pasquale, "Cognition-Enhancing Drugs: Can We Say No?," *Bulletin of Science, Technology, and Society* 30, no. 1 (2010): 9–13; John

Patrick Leary, "Keywords for the Age of Austerity 19: Resilience," *Keywords: The New Language of Capitalism* (blog), June 23, 2015, https://keywordsforcapitalism.com/2015/06/23/keywords-for-the-age-of-austerity-19-resilience/.

61. Jonathan Crary, *24 / 7: Late Capitalism and the Ends of Sleep* (London: Verso, 2013).

62. Ray Kurzweil, *The Age of Spiritual Machines: When Computers Exceed Human Intelligence* (New York: Penguin, 1999), 166.

63. Sean Dorrance Kelly, "A Philosopher Argues That an AI Can't Be an Artist: Creativity Is, and Always Will Be, a Human Endeavor," *MIT Technology Review,* February 21, 2019, https://www.technologyreview.com/s/612913/a-philosopher-argues-that-an-ai-can-never-be-an-artist/.

64. For a description of the many flaws in the Turing test, see Brett Frischmann and Evan Selinger, *Re-engineering Humanity.*

65. Margaret Boden, *AI: Its Nature and Future* (Oxford: Oxford University Press, 2016), 119. The same goes for emotion, as we explored "caring" robots in Chapter 2.

66. Gabe Cohn, "AI Art at Christie's Sells for $432,500," *New York Times*, October 25, 2018, https://www.nytimes.com/2018/10/25/arts/design/ai-art-sold-christies.html.

67. Tyler Sonnemaker, "No, an Artificial Intelligence Can't Legally Invent Something—Only 'Natural Persons' Can, Says US Patent Office," *Business Insider*, April 29, 2020, https://www.businessinsider.com/artificial-inteligence-cant-legally-named-inventor-us-patent-office-ruling-2020-4.

68. Korakrit Arunanondchai, *With History in a Room Filled with People with Funny Names 4*, 2018, DCP, 24:00; Adriana Blidaru, "How to Find Beauty in a Sea of Data: Korakrit Arunanondchai at CLEARING," *Brooklyn Rail*, May 2017, https://brooklynrail.org/2017/05/film/How-to-find-beauty-in-a-sea-of-data-Korakrit-Arunanondchai-at-C-L-E-A-R-I-N-G.

69. Lawrence Joseph, "Visions of Labour," *London Review of Books* 37, no. 12 (June 18, 2015), https://www.lrb.co.uk/v37/n12/lawrence-joseph/visions-of-labour. I wish to thank Lawrence Joseph for permission to excerpt "Visions of Labour."

70. Anthony Domestico, "An Heir to Both Stevens and Pound," *dotCommonweal,* October 20, 2015, https://www.commonwealmagazine.org/blog/heir-both-stevens-pound.

71. Chris Arnold, "Despite Proven Technology, Attempts to Make Table Saws Safer Drag on," NPR, August 10, 2017, https://www.npr.org/2017/08/10 /542474093/despite-proven-technology-attempts-to-make-table-saws -safer-drag-on.

72. W. Brian Arthur, "Increasing Returns and the New World of Business," *Harvard Business Review,* July–August 1996, https://hbr.org/1996/07 /increasing-returns-and-the-new-world-of-business.

73. Ken Johnson, "Review: Simon Denny Sees the Dark Side of Technology at MOMA," *New York Times,* May 28, 2015, https://www.nytimes.com/2015 /05/29/arts/design/review-simon-denny-sees-the-dark-side-of-technology -at-moma-ps1.html.

74. Ibid.

75. As Rosalind Krauss observed of Sol LeWitt's endlessly repeated, math-ematically determined grids and boxes, "LeWitt's outpouring of example, his piling up of instance, is riddled with system, shot through with order. There is, in *Variations of Incomplete Open Cubes,* as they say, a method in this madness. For what we find is the 'system' of compulsion, of the obsessional's unwavering ritual, with its precision, its neatness, its finicky exactitude, covering over an abyss of irrationality. It is in that sense design without reason, design spinning out of control." Rosalind Krauss, "LeWitt in Progress," *October* 6 (1978): 46, 56.

76. William E. Connolly, *The Fragility of Things: Self-Organizing Processes, Neoliberal Fantasies, and Democratic Activism* (Durham, NC: Duke University Press, 2013).

77. For example, Philip N. Meyer, "The Darkness Visible: Litigation Stories & Lawrence Joseph's *Lawyerland,*" *Syracuse Law Review* 53, no. 4 (2003): 1311.

78. Roger Cotterrell, *Law, Culture and Society: Legal Ideas in the Mirror of Social Theory* (London: Routledge, 2006): 15.

79. Richard H. Brown, *A Poetic for Sociology: Toward a Logic of Discovery for the Human Sciences* (New York: Cambridge University Press, 1977), 26.

80. I draw this terminology from Jurgen Habermas, *The Theory of Communi-cative Action*, vol. 2, *System and Lifeworld* (Cambridge, UK: Polity Press, 1985). The three realms of subjective, objective, and intersubjective map (if imperfectly) to understandings of beauty, truth, and justice.

81. See "Critical Algorithm Studies: A Reading List," updated December 15, 2016, https://socialmediacollective.org/reading-lists/critical-algorithm

-studies/. For a recent critique of behaviorism, see Frischmann and Selinger, *Re-engineering Humanity.*

82. Rob Kitchin, *The Data Revolution* (London: Sage, 2014), 2 ("Strictly speaking, then, this book should have been called *The Capta Revolution.*"). As the photographer Edward Steichen argued, "Every photograph is a fake, from start to finish, a purely impersonal, unmanipulated photograph being a practical impossibility." Edward J. Steichen, "Ye Fakers," *Camera Work* 1 (1903): 48. See also Wallace Stevens, "The Man with the Blue Guitar," in *Collected Poetry and Prose,* eds. Joan Richardson and Frank Kermode (New York: Penguin, 1997).

83. In *The Language Animal: The Full Shape of the Human Linguistic Capacity* (Cambridge, MA: Harvard University Press, 2014), Charles Taylor distinguishes designative and constitutive views of language, and decisively opts for the latter as a fuller description of the "human linguistic capacity."

84. Taylor, *The Language Animal*, 24. The concept of resonance is formally elaborated and rigorously explored in Hartmut Rosa, *Resonance: A Sociology of the Relationship to the World* (London: Polity Press, 2019). Taylor further observes that "this kind of change is analogous, on a more abstract and objectified level, to our changing our mode of scientific enquiry by shifting paradigms." Taylor, *The Language Animal*, 24–25. He continues: "Self-understanding, and human understanding in general, can also be enhanced by coming to recognize new models; and that is why literature is such a source of insight. . . . Humboldt sees us as pushed . . . to open up to speech areas which were previously ineffable. Certainly poets are embarked on this enterprise: T. S. Eliot speaks of 'raids on the inarticulate.' Humboldt, for his part, posits a drive [*Trieb*] 'to couple everything felt by the soul [mind] with a sound.'"

85. Robert J. Shiller, *Narrative Economics: How Stories Go Viral and Drive Major Economic Events* (Princeton: Princeton University Press, 2019).

86. Steve LeVine, "The Economist Who Wants to Ditch Math," *Marker: Medium*, November 5, 2019, https://marker.medium.com/robert-shiller -says-economics-needs-to-go-viral-to-save-itself-f157eceb4c7d.

87. Jens Beckert and Richard Bronk, eds., *Uncertain Futures: Imaginaries, Narratives, and Calculation in the Economy* (Oxford: Oxford University Press, 2018).

88. J. M. Balkin, *Cultural Software: A Theory of Ideology* (New Haven: Yale University Press, 1998), 479.

89. Frank Pasquale, Lenore Palladino, Martha T. McCluskey, John D. Haskell, Jedidiah Kroncke, James K. Moudud, Raúl Carrillo et al., "Eleven Things They Don't Tell You about Law & Economics: An Informal Introduction to Political Law and Economics," *Law & Inequality: A Journal of Theory and Practice* 37 (2019): 97–147.

90. Maurice Stucke and Ariel Ezrachi, *Competition Overdose: How Free Market Mythology Transformed Us from Citizen Kings to Market Servants* (New York: HarperCollins, 2020).

91. Mireille Hildebrandt, "A Vision of Ambient Law," in *Regulating Technologies,* eds. Roger Brownsword and Karin Yeung (Portland, OR: Hart Publishing, 2008), 175.

92. Catherine Despont, "Symbology of the Line," in Ernesto Caivano, *Settlements, Selected Works 2002–2013* (New York: Pioneer Works Gallery, 2013). "Ostensibly, Caivano's work recounts an epic tale in which two lovers, separated by a quest, are forced to reconnect across fluctuating time and multiple dimensions" (31).

93. Ernesto Caivano, *Settlements, Selected Works 2002–2013* (New York: Pioneer Works Gallery, 2013).

94. David Edward, *Artscience: Creativity in the Post-Google Generation* (Cambridge, MA: Harvard University Press, 2010).

95. Ernesto Caivano, *Settlements*, 40–53.

96. Hartmut Rosa, *The Social Acceleration of Time* (New York: Columbia University Press, 2013).

97. Brian J. Robertson, *Holacracy: The New Management System for a Rapidly Changing World* (New York: Holt, 2015).

98. Katherine Barr, "AI Is Getting Good Enough to Delegate the Work It Can't Do," *Harvard Business Review,* May 12, 2015, https://hbr.org/2015/05/ai-is-getting-good-enough-to-delegate-the-work-it-cant-do.

99. Alex Rosenblat and Luke Stark, "Algorithmic Labor and Information Asymmetries: A Case Study of Uber's Drivers," *International Journal of Communication* 10 (2016): 3758–3784; but see also Alison Griswold, "Uber Drivers Fired in New York Can Now Appeal Before a Panel of Their Peers," *Quartz,* November 23, 2016, https://qz.com/843967/uber-drivers-fired-in-new-york-can-now-appeal-before-a-panel-of-their-peers/.

100. For a description and critical commentary, see the opening of Nick Dyer-Whiteford, *Cyber-Proletariat: Global Labour in the Digital Vortex* (Chicago: University of Chicago Press, 2015): 1–4.

101. Illah Reza Nourbakhsh, *Robot Futures* (Cambridge, MA: MIT Press, 2013), xix-xx.

102. Simone Weil, *Gravity and Grace,* trans. Arthur Wills (New York: Putnam, 1952), 10.

103. *Walter Benjamin: Selected Writings*, eds. Howard Eiland and Michael W. Jennings, trans. Edmund Jephcott, vol. 4, *1938-1940* (Cambridge, MA: Belknap Press, 2006), 392. Walter Benjamin, in the ninth thesis of his 1940 essay "Theses on the Philosophy of History," wrote about artist Paul Klee's *Angelus Novus*: "This is how the angel of history must look . . . [and a] storm drives him irresistibly into the future to which his back is turned, while the pile of debris before him grows toward the. What we call progress is *this* storm." For a discussion of Benjamin's interest in Klee's *Angelus Novus*, see Enzo Traverso, *Left-Wing Melancholia: Marxism, History, and Memory* (New York: Columbia University Press, 2017), 178–181.

104. Masahiro Mori, "The Uncanny Valley: The Original Essay by Masahiro Mori," trans. Karl F. MacDorman and Norri Kageki, *IEEE Spectrum*, June 12, 2012, https://spectrum.ieee.org/automaton/robotics/humanoids/the-uncanny-valley.

105. Teiji Furuhashi's installation *Lovers* is a computer controlled, five-channel laser disc / sound installation with five projectors, two sound systems, two slide projectors, and slides (color, sound). For a description and video of the Museum of Modern Art's 2016–2017 exhibition of the work, see "Teiji Furuhashi: Lovers," MoMA, 2016, https://www.moma.org/calendar/exhibitions/1652.

ACKNOWLEDGMENTS

New Laws of Robotics is a project that flowed naturally from *Black Box Society*, my last book with Harvard University Press. While *Black Box* was primarily a work of critique, *New Laws* offers a vision of how technology can be better integrated into society. I wish to thank Thomas LeBien for his wise advice and expert editing tips while he was at Harvard University Press, and I am very grateful to James Brandt for the same and for shepherding the project to completion.

With respect to the epigraphs: I wish to thank Lawrence Joseph for permission to quote a stanza of "In Parentheses," from *A Certain Clarity: Selected Poems* (New York: Farrar, Straus and Giroux, 2020). I am grateful to be able to include an excerpt from *Between Past and Future* by Hannah Arendt, copyright © 1954, 1956, 1957, 1958, 1960, 1961, 1963, 1967, 1968 by Hannah Arendt. Used by permission of Viking Books, an imprint of Penguin Publishing Group, a division of Penguin Random House LLC. All rights reserved.

I am fortunate to have had so many thoughtful and caring colleagues and friends who selflessly offered their expertise, challenged my assumptions, and discussed the many facets of robotics and AI as I worked on *New Laws*. I completed the book while teaching and researching at the University of Maryland. Bruce Jarrell, Phoebe Haddon, and Donald Tobin were supportive of my work, and I am grateful for their vision. Maryland's law school and its *Journal of Health Care Law and Policy* sponsored a workshop on medical automation and robotics law and policy which I convened, bringing together technologists, academics, and policy experts.

Friends like Diane Hoffmann, Danielle Keats Citron, Don Gifford, Bob Condlin, and Will Moon made Maryland an academic home for me. Sue McCarty and Jennifer Elisa Chapman at the library were exceptionally dedicated and helpful. I am sorry to leave Maryland so near the completion of this project—I would love to celebrate it there. But I look forward to working with my new colleagues at Brooklyn Law School in 2020. They were extraordinarily welcoming and insightful during my visit there in 2019.

In 2016, Jack Balkin, Yale Law School, and the Yale Information Society Project offered me an invaluable opportunity to co-organize the conference "Unlocking the Black Box: The Promise and Limits of Algorithmic Accountability in the Professions." The conference featured work from many scholars exploring and promoting algorithmic accountability. Many demonstrated the critical role of labor in shaping complex technical systems. The conference, as well as subsequent visits to the Yale ISP, enriched this work tremendously.

I am also grateful for the opportunity to present drafts of chapters of this book at several academic institutions. In Asia, scholars at the Research Institute of the Social Sciences and Humanities at National Taiwan University, Academia Sinica, National Cheng Kung University, and Hong Kong University taught me much about the ethical, legal, and social implications of AI. In Canada, Queens University and the University of Toronto generously convened interdisciplinary workshops on AI and law that I attended, and Humber College invited me to present on computational evaluations of persons. Researchers at the University of Western Australia, the University of Sydney (including its the Social Sciences and Humanities Advanced Research Centre), the University of Melbourne, Monash University, and the Queensland University of Technology were also generous in commenting freely on ideas presented from the book during my visits to Australia. I look forward to continuing this work in an ongoing research project on automated decision-making there. In Europe, experts convened at the Free University of Brussels, the London School of Economics, the European University Institute, the Pontifical Academy of the Social Sciences, Manchester University, the Center for Research in Arts, Social Sciences, and Humanities at Cambridge University, and Edinburgh University generously listened to chapters or sections of the book, and offered very helpful critiques.

I am grateful to the many North American law schools that hosted me to speak on my work in AI law and policy, including Northeastern, the University of San Diego, Yale, Columbia, Harvard, the University of Pennsylvania, Seton Hall, Washington University, Georgetown, Osgoode Hall, Fordham, Case Western, Ohio State, New York University, Temple, and Rutgers-Camden. Presentations at other university divisions also helped me sharpen my critiques of automation. These included Columbia University's School of International and Public Affairs, Department of History, and Division of Social Science; the Berman Institute of Bioethics at Johns Hopkins University, the Hariri Institute for Computing and Computational Science and Engineering at Boston University, the University of Virginia's School of Engineering and Applied Science and Institute for Advanced Studies in Culture, and the Princeton Program in Law and Public Affairs.

Policymakers have also taken interest in my approach, and it was an honor to receive their feedback. I presented material on the automated public sphere to the Media Authority of Berlin and Brandenburg and to representatives of Directorates-General of the European Commission. Staff from the European

Medicines Agency, Center for Medicare and Medicaid Innovation, National Committee on Vital and Health Statistics, and Food and Drug Administration were helpful interlocutors as I developed my approach to medical automation and robotics law and policy. Testifying on data policy and algorithms before the House Energy and Commerce Committee and the Senate Committee on Banking, Housing, and Urban Affairs was a good opportunity to present my critique of "machines judging humans" and to learn from staffers. Thoughtful interlocutors from the Federal Trade Commission, Office of the Privacy Commissioner of Canada, and the Maryland Administrative Law Judiciary also generously listened to, and commented on, presentations of parts of chapters in this work.

Civil society groups and NGOs also helped shape this book. I am immensely grateful to the AI Now Institute, Data & Society, the Association for the Promotion of Political Economy and Law, the Hoover Institution, the American Society of International Law, the Social Science Research Council, the Minderoo Foundation, the Modern Money Network, the Royal Society for the Encouragement of Arts, Manufactures and Commerce, the American Society for Law, Medicine, and Ethics, the Nobel Prize Dialogue, the Friedrich Ebert Stiftung, Re:Publica, the Rosa Luxemburg Foundation, the Manchester Co-op, g0v, the Interdisciplinary Perspectives on Accounting Conference, and the Edinburgh Futures Institute for giving me forums to present my research to open, diverse audiences beyond the ken of universities, governments, and corporations.

I also wish to thank all of the following for their comments on presentations or work that informed this volume: Kamel Ajji, Mark Andrejevic, Susan Bandes, Christopher Beauchamp, Benedetta Brevini, Raúl Carrillo, Clement Chen, Hung-Ju Chen, Wentsong Chiou, Julie Cohen, Nicole Dewandre, David Golumbia, Karen Gregory, John Haskell, Susan Herman, Amanda Jaret, Kristin Johnson, Ching-Yi Liu, Alice Marwick, Martha McCluskey, John Naughton, Julia Powles, Evan Selinger, Alicia Solow-Niederman, Simon Stern, and Ari Ezra Waldman.

Finally, I acknowledge and celebrate the constant support of my partner, Ray. From the surfeit of books and papers filling our apartment, to my constant trips to conferences, to the creep of work into late hours and vacations, he has put up with a lot, always with good grace. There's no one I'd rather face the good and bad times with, the *grandeur et misère* of being human, than him.

INDEX

Brazil, 97, 122
Brennan-Marquez, Kiel, 129
Brexit, 22, 89
Brooks, Rosa, 169
Broussard, Meredith, 212
Brown, Richard H., 223–224
Brynjolfsson, Erik, 102, 104
Buolamwini, Joy, 125
Burton, Robert A., 161
Bush, George W., 293n95

Cadwalladr, Carole, 95
Caivano, Ernesto, 202, 226–227, 309n92; *After the Woods*, 226; *Echo Series*, 226; *Philapores Navigating the Log and Code*, 227
California, 62, 74, 83, 110
Canada,16
cancer, 18, 23, 35–37, 41–42
Capers, Bennett, 123
caregiving, 28, 49–55, 191. *See also* child care; elder care
Carney, Mark, 26
Carpenter, Julie, 15
cars, 6; automation of, 19–20. *See also* self-driving cars
CCP. *See* Chinese Communist Party
CCTV cameras, 122
CDSS. *See* clinical decision support software
censorship, 99, 101, 103, 160, 166, 205
Centers for Medicare and Medicaid Services, 44
Chamayou, Grégoire, 152–153
charter schools (online or virtual), 16, 83–84
chatbots, 8, 12, 47, 88, 219, 248n50
Chen, Yongxi, 280n71
Cheung, Anne S. Y., 280n71
Chiang Kai-Shek, 163
child care, 56, 70, 71, 174. *See also* caregiving

children: dangers of behaviorist approach to educating, 75–77; and the downsides of online learning, 83, 86; ethics of collecting data from, 73–75; importance of human teachers for, 25, 65, 68–71; in less developed countries, 81–83; and positive robotic helpers, 77–81, 87; vulnerable to misuse of online media, 72, 97
China: AI arms race in, 165; social credit scoring in, 10–11, 132; development programs in, 169; edtech in, 60–61, 75, 77; as exporter of surveillance equipment, 166–167; facial recognition software in, 126; health sector in, 191; human-robot interaction in, 54–55; and land mines, 157; loyalty assessment in, 11, 160; military and military spending of, 147, 154, 159, 167–168, 291n74, 292n84; Muslims in, 126, 160, 167; online media platforms in, 60, 115; penalties in, for political involvement, 136; politics in, 160; prisons in, 124; and Project Dragonfly, 166; social credit in, 136–139, 160; surveillance in, 61, 160, 167–168; and Taiwan, 160, 163–164; traditional classrooms in, 81; treatment of the Uyhgurs in, 122
Chinese Communist Party (CCP), 160, 161
Christie's (auction house), 220
CIA, 165
Cifu, Adam, 58
Citibank, 119, 136
civics, 65, 82, 91
Civilian Conservation Corps, 186
civil liberties, 128
civil rights, 122
clean energy, 182

cleaning (jobs in), 5, 49, 182; cleaners, robotic / automated, 171, 172
click farms, 92, 108
climate change, 169, 175, 186, 206; denialists, 98; and the shift from manufacturing to service jobs, 190; and transport solutions, 179
clinical decision support software (CDSS), 37–38, 244n16, 244n16, 244–245n17
Clinton, Hillary, 90, 93, 94, 98
"coding boot camps," 66
Cohen, Julie, 276n29
Cole, August, 147
collective bargaining, 88, 177. *See also* unions
colleges and universities: aims of, 63–64; for caregivers, 192; citizens entitled to four years of, 175; content of (intrinsic vs. instrumental), 178–179; costs of, 85, 87; evaluation in, 141–142; for-profit, 134; free, 135; making students "robot-proof," 173–174; online, 84–86; STEM training at, 173. *See also* student debt
conspiracy theories, 94, 98, 113
consumer protection, 15, 58, 201; and deceptive marketing, 102
contact tracing, 11
content moderation, 91, 95, 96, 107, 117–118
copyright infringement, 165
coronaviruses. *See* COVID-19 pandemic; pandemics
"cost disease," 171–172, 186–190, 196, 198, 299n68
COVID-19 pandemic, 100, 166, 191, 194; and data deficiencies, 243n3; and essential workers, 172–173; preparedness for, 159–160, 184–185; and recovery of jobs, 185–186, 195; and universities, 62

Crary, Jonathan, 217
credit and credit scoring, 10, 30, 131–136, 140, 282n83. *See also* social credit
Criado Perez, Caroline, 39
crime / criminality, 18, 40, 143; and facial recognition, 10, 125, 128–130; future, 17, 123; and the right to be forgotten, 106
critical race theory, 212
Crootof, Rebecca, 289n60
cryptocurrency, 194, 205, 206
customer service. *See* service, jobs in
cyberattacks, 10
cyberlibertarianism, 18–19, 99, 105
cybernetics, 145–146
cybersecurity, 158
cyberwarfare, 30, 31

Damiano, Luisa, 215
Darling, Kate, 79–80
data: in art creation, 220; biases in, 30, 38–39, 98–99, 124–126; in education, 62, 72–77, 83, 85; ethics of collecting, and limits on, 16, 18–19, 72–75, 120; etymology of the term, 224; in health care, 35–38, 41–46; misuse of, 101–103, 127, 128, 130, 132, 133, 135, 140, 141, 167; and privacy, 106, 127–128, 201; proper collection, storage, and management of, 15, 33, 57, 107, 174, 192; protection laws, 201, 205; value of, 13, 80, 165, 179, 182
Davies, Will, 24, 142
debt, 30, 131–136; and default, 280–281n71. *See also* lending; student debt
decryption, 10, 206
deep fakes, 118, 164, 269–270n71
deep learning, 39
defamation, 93, 98, 106, 109
Defense Advanced Research Projects Agency, 147

encryption, 10, 205
Engelbart, Doug, 237n29
environment, 171, 185. *See also*
    climate change
"essential workers," 172–173
ethics, 14, 16, 19, 29, 51, 80, 135, 158,
    201, 213, 214, 230, 234n8; and the
    arts and humanities, 200; and "bot
    disclosure," 7–8; and care, for inter-
    active robots, 80; in education, 70,
    72; and judgmental AI, 139–140;
    and military tech, including killer
    robots, 9–10, 17, 146, 149–153, 158,
    158, 165–166, 168; and online media,
    94–95; professional, as regards
    robotization, 14, 187–188; and the
    right to be forgotten, 106–107; and
    robotization of policing, 123–131,
    159; and self-driving cars, 22;
    and surveillance, 131; and the
    tension between pragmatists and
    futurists, 28
Europe: and ethics for AI, 234n8; and
    federalism, 176; privacy law in, 5;
    regulators in, 16; and the "right
    to an explanation," 18; rights of
    erasure in, 74; and the right to be
    forgotten, 91, 105–107; robotic care
    in, 55, 56; traditional classrooms in,
    81. *See also specific countries*
European Union. *See* Europe
exculpatory clauses, 40, 41
Ex Machina (2015 film), 202–203, 204,
    205, 213
extremism, 90, 98, 104, 108, 117
Eyal, Gil, 23, 239n54

Facebook, 214; algorithms, 181, 225;
    Basics, rejected in India, 82–83;
    "cleaners," 97; disruption of tradi-
    tional media by, 30, 112, 117; goal
    of, to ensure users click on ads, 68,
    90; hate speech, fake news, and

propaganda on, 92–95, 103, 114;
    like / not-like binaries of, 47; and
    reporting of teen suicidality, 73;
    restoring responsibility and
    professionalism to, 91, 98, 100,
    113–115
facial analysis, 18, 72, 74–77, 79, 85,
    123, 128–131, 140, 214
facial recognition (FR), 1, 10, 36,
    60–61, 119, 124–128, 131, 148, 151,
    171, 281n74; in China, 138, 160;
    and persons of color, 123, 125–127,
    245n19
fact checking, 95, 113–114
"fake news," 30, 90, 93, 94, 103,
    113–114, 159. *See also* propaganda
farming. *See* agriculture
Farrell, Henry, 161
FDA. *See* US Food and Drug
    Administration
Feenberg, Andrew, 176
feminism, 212
Ferguson, Cat, 101
finance: AI in, 12, 131–136, 206, 228;
    and the drive for power and domi-
    nance, 10,142, 143, 201, 229; the
    renewal of public, 162, 192–196.
    See also financial technology
financial technology (fintech), 120,
    132, 133, 139
Fineman, Martha, 55
Finland, 174
First Amendment. *See* US Constitution:
    First Amendment
Fitbits, 72
Florida, 102
Ford, 186
Forster, E. M., 196–197, 216
France, 159, 174, 249n63
Frank, Robert, 143
Frase, Peter, 300n2
freedom of speech / expression, 93, 106,
    107, 108–110, 127, 269n68

Free Speech for People, 269n68
Freidson, Eliot, 57
Furuhashi, Teiji, *Lovers*, 231, 310n105
Future of Life Institute, 147
futurism, 28; medical, 33, 242n1; and policing, 121–122; and science fiction, 300n2. *See also* Afrofuturism

gait recognition, 138, 140, 148
Gal, Danit, 54–55
Galatea 2.2 (Powers), 209
Gates Foundation, 61
Gaycken, Sandro, 165
gays and lesbians, and facial analysis, 128
Geertz, Clifford, 201
gender bias. *See* discrimination: sexism
gender gap: in data, 39; in technology, 85
General Dynamics, 165
Geneva Conventions, 151
genomics, 245n21
Georgia Tech University, 61
Germany, 113, 159, 177, 180
global warming. *See* climate change
Godsey, Michael, 261n85
Goel, Ashok, 61–62
Goffman, Erving, 73–74
gold standard, 74, 193, 194
Google, 36, 117, 218, 241n69; advertising on, 68, 215; algorithms, 181, 225; Android Market, 45; Android operating system, 72; and anti-Semitism, 95; Assistants, 8, 110–112; Glass, 6, 126; and Holocaust denial, 108; manipulation of searches on, 101–102, 212; Maven, 164–165; and military tech collaboration, 164–166; Maps, 102; medical searches on, 34–35; Project Dragonfly, 166; propaganda and misinformation on, 93–95, 97, 98–100, 115; racist search results on, 212; and the

rights of those being searched, 105–107
Gove, Michael, 22
Goyal, Nikil, 67
Gray, Erik, 209
Great Britain. *See* United Kingdom
Great Depression, 186, 195
Great Recession, 62
Green, Ben, 179
Green New Deal, 195
Grimmelmann, James, 109

Habermas, Jurgen, 109, 140, 282n85
hacking / hackers, 12, 74, 96, 104, 108, 213; and arms races, 154, 155, 167, 206
Hale, Robert Lee, 176
Hanwang (Chinese tech firm), 60–61
harassment, online, 12, 30, 95
Hartzog, Woodrow, 126
Harvard Business Review, 227
Harvey Mudd College, 85
hate speech, 12, 18, 30, 92, 103, 108, 113, 118, 165
Hawkins, Amy, 126
Head, Simon, 43
health care, 2, 29, 31, 33–34, 188, 189, 190, 199, 202, 208; accountability / liability in, 39–41; and caregiving, 49–55; and diagnostics / pattern recognition, 35–38; the human touch in, 55–59; and "learning health care systems," 41–45; and medical searches, 34–35; and racial disparities, 38–39; spending on, 43–44, 190–192, 247n45; and symptom-analyzing apps, 4, 120; and therapy apps, 45–49
Her (2013 film), 202, 203–204, 213, 214
higher education. *See* colleges and universities
high frequency trading, 11, 20

Khan Academy, 84
"killer robots." *See* weapons: robots as
King, Thomas, 112
Klee, Paul, 310n103
Knight Capital, 155
Korte, Travis, 112
Krauss, Rosalind, 222, 307n75
Kurzweil, Ray, 218

labor. *See* unions; workers
Lakoff, George, 212–213
land mines, 156–157, 289n51
law enforcement, 17–18, 102, 122–124,
    131, 165, 199. *See also* facial
    recognition; policing
lawfare, 31
LAWS. *See* lethal autonomous
    weapons systems
laws of robotics, Asimov's, 2–3, 233n6
laws of robotics, new, 3–13, 90, 117–118,
    170, 171, 185, 213, 225, 229, 233n6,
    234n8; first, 29, 33, 91, 100, 171,
    177; fourth, 109, 118, 146, 158, 220;
    second, 110, 118, 146; third, 10, 31,
    91, 101, 136, 143–144, 171, 217, 225
Lego Mindstorms, 87
lending, 131–136; and loan forgiveness,
    279n58; micro-, 132, 134; predatory,
    133, 134, 135
lethal autonomous weapons systems
    (LAWS), 9–10, 11, 31, 92, 148,
    155–159, 162, 287n41
Levandowski, Anthony, 179
Levine, Yasha, 165
LeWitt, Sol, 307n75
liability: in health care, 37, 39–41: in
    the military, 157. *See also* defama-
    tion; insurance; malpractice
libertarianism, 105, 184. *See also*
    cyberlibertarianism
licensing, for technology, 12, 22, 58,
    120, 127

licensing, in health care, 49, 57, 58, 88,
    187
Lipson, Hod, 18
Liu, Cixin, 209
logistics (jobs in), 5, 182, 192
Lomas, Natasha, 112
Lopez, Matthew, 216
L'Oreal, 119

machine learning, 5, 7, 10, 39, 142,
    201, 216, 225, 226; accountability
    and transparency in, 120–121, 174;
    in art creation, 219; best uses of, 64,
    149; and children, 73, 75; and cred-
    itworthiness, 132; and criminality,
    18; and diversity, 22; and drone
    footage, 164; and facial / number /
    pattern recognition, 36, 128–129;
    and manipulation of search results,
    101–102; and medical research, 39;
    and obscenity, 104; and racism /
    sexism, 99; and the service sector,
    197. *See also* algorithms; data
Madrigal, Alexis, 94
malpractice, 37, 38, 44. *See also*
    liability: in health care
management consultants, 2, 222
manga, 54
manufacturing, 8, 190, 192, 197;
    transformed by automation, 14, 26,
    83, 186. *See also* assembly lines
Marlowe, Christopher, *Faust,* 230
Mars (company), 119
MarsCat (robotic pet), 215–216
Martin, Agnes, 226
maskirovka (military deceptions),
    150
Massachusetts Institute of Technology
    (MIT), 13, 46–47, 51, 53, 78, 79
Mayo Clinic, 35
Mazzucato, Mariana, 193
McAfee, Andrew, 102, 104